STUDY GUIDE for Fourth Edition of ESSENTIALS OF FIRE FIGHTING

Developed by Cindy Pickering
Edited by Barbara Adams

Published by
Fire Protection Publications Oklahoma State University Stillwater, Oklahoma

ISBN 0-87939-146-4

Second Edition

Printed in the United States of America

10

Oklahoma State University in compliance with Title VI of the Civil Rights Act of 1964 and Title IX of the Educational Amendments of 1972 (Higher Education Act) does not discriminate on the basis of race, color, national origin or sex in any of its policies, practices or procedures. This provision includes but is not limited to admissions, employment, financial aid and educational services.

Dedication

This manual is dedicated to the members of that unselfish organization of men and women who hold devotion to duty above personal risk, who count on sincerity of service above personal comfort and convenience, who strive unceasingly to find better ways of protecting the lives, homes and property of their fellow citizens from the ravages of fire and other disasters . . . ***The Firefighters of All Nations.***

Dear Firefighter:

The International Fire Service Training Association (IFSTA) is an organization that exists for the purpose of serving firefighters' training needs. Fire Protection Publications is the publisher of IFSTA materials. Fire Protection Publications staff members participate in the National Fire Protection Association and the International Association of Fire Chiefs.

If you need additional information concerning our organization or assistance with manual orders, contact:

Customer Services
Fire Protection Publications
Oklahoma State University
930 N. Willis
Stillwater, OK 74078-8045
1 (800) 654-4055 Fax: (405)744-8204

For assistance with training materials, recommended material for inclusion in a manual, or questions on manual content, contact:

Editorial Department
Fire Protection Publications
Oklahoma State University
930 N. Willis
Stillwater, OK 74078-8045
(405) 744-5723 Fax: (405) 744-4112
E-mail: editors@osufpp.org

NOTICE

The questions in this study guide are taken from the information presented in the fourth edition of ***Essentials of Fire Fighting,*** an IFSTA-validated manual. The questions are *not validated test questions and are not intended to be duplicated or used for certification or promotional examinations;* this guide is intended to be used as a tool for studying the information presented in ***Essentials of Fire Fighting.***

Table of Contents

Preface vi
How to Use this Book vii
Chapters
1 Firefighter Orientation and Safety 3
2 Fire Behavior 23
3 Building Construction 43
4 Firefighter Personal Protective Equipment 55
5 Portable Extinguishers 75
6 Ropes and Knots 89
7 Rescue and Extrication 103
8 Forcible Entry 127
9 Ground Ladders 151
10 Ventilation 169
11 Water Supply 183
12 Fire Hose 199
13 Fire Streams 229
14 Fire Control 249
15 Fire Detection, Alarm, and Suppression Systems 269
16 Loss Control 289
17 Protecting Evidence for Fire Cause Determination 297
18 Fire Department Communications 309
19 Fire Prevention and Public Fire Education 321

Answers to Chapters
1 341
2 345
3 347
4 351
5 353
6 355
7 357
8 361
9 365
10 367
11 369
12 373
13 379
14 383
15 387
16 391
17 393
18 397
19 401

Preface

This study guide is designed to help the reader understand and remember the material presented in **Essentials of Fire Fighting**, fourth edition. It identifies important information and concepts from each chapter and provides questions to help the reader study and retain this information. In addition, the study guide serves as an excellent resource for individuals preparing for certification or promotional examinations.

Much time and effort go into the design, development, layout, and printing of any publication. This **Study Guide for the Fourth Edition of Essentials of Fire Fighting** is no exception. I would like to give special recognition to Keith Flood; Ansul, Inc.; Conoco, Inc.; Hale Fire Pump Co., Inc.; Elkhart Brass Manufacturing Company; and Mount Shasta (CA) Fire Protection District for contributing photographs used in this publication.

I would also like to extend a special thank-you to the following members of the Fire Protection Publications staff whose contributions made possible the technical accuracy and visual appeal of this publication.

Barbara Adams, Associate Editor
Susan S. Walker, Instructional Development Coordinator
Gordon Earhart, Senior Publications Editor
Don Davis, Publications Production Coordinator
Desa Porter, Senior Graphic Designer
Ann Moffat, Graphic Design Analyst
Dean Clark, Temporary Senior Graphic Designer
Brent Meisenheimer, Research Technician
Dustin Stokes, Research Technician

Cindy Pickering
Curriculum Specialist

How to Use this Book

This study guide is developed to be used in conjunction with and as a supplement to the fourth edition of the IFSTA manual, **Essentials of Fire Fighting.** The guide has been separated into Firefighter I and Firefighter II sections to make the study process applicable to the various training entities that train to these levels. Most chapters contain material relevant to both Firefighter I and Firefighter II training. The training level is indicated at the beginning of each unit and on tabs at the edge of each page.

The questions in this guide are designed to help you remember information and to make you think—they are *not* intended to trick or mislead you. To derive the maximum learning experience from these materials, use the following procedure:

Step 1: Read one chapter at a time in the **Essentials of Fire Fighting** manual. After reading the chapter, underline or highlight important terms, topics, and subject matter in that chapter.

Step 2: Open the study guide to the corresponding chapter. Answer all of the questions in the study guide for that chapter. You may have to refer to a dictionary or the glossary in the **Fire Service Orientation and Terminology (O&T)** manual for terms that appear in context but are not defined. After you have defined terms and answered all questions possible, check your answers with those in the answer section at the end of the study guide.

✔**Note:** *Do not* answer each question and then immediately check the answer for the correct response.

If you find that you have answered any question incorrectly, find the explanation of the answer in the **Essentials of Fire Fighting** manual. The number in parentheses after each answer in the answer section identifies the page on which the answer or term can be found. Correct any incorrect answers, and review material that was answered incorrectly.

Step 3: Go to the next chapter of the manual and repeat Steps 1 and 2.

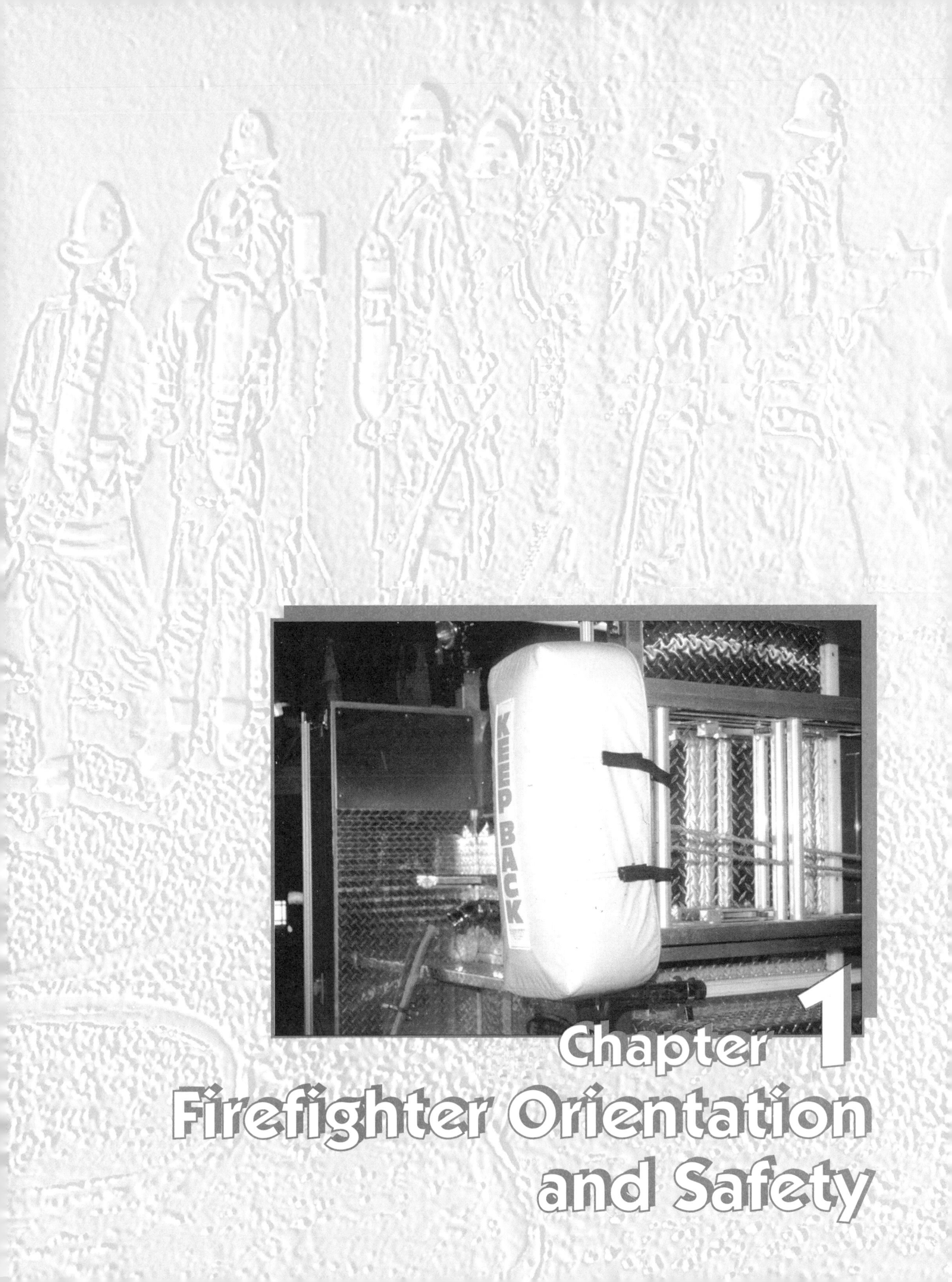

Chapter 1
Firefighter Orientation and Safety

Chapter 1 Firefighter Orientation and Safety

FIREFIGHTER I

Matching

A. Match to their definitions terms associated with orientation and safety. Write the appropriate letters on the blanks.

____ 1. Diagram showing the structure of the fire department and its chain of command

____ 2. Principle that a person can only report to one supervisor

____ 3. Pathway of responsibility from the highest level of the department to the lowest

____ 4. Number of personnel that one individual can effectively manage

____ 5. Setting the limits or boundaries for expected performance and enforcing them

____ 6. Standard operating unit of a fire department

____ 7. Guide to decision making within an organization

____ 8. Detailed guide to action

____ 9. Written or verbal instruction based on a policy or procedure

____ 10. Standard set of actions at the core of every fire fighting incident plan

____ 11. Method of sorting victims by the severity of their injuries

a. Chain of command
b. Directive
c. Company
d. Unity of command
e. Order
f. Span of control
g. Organizational chart
h. Triage
i. Policy
j. Discipline
k. SOPs
l. Procedure

True/False

B. Write *True* or *False* before each of the following statements. Correct those statements that are false.

_____ 1. New firefighters choose between two categories: career or volunteer.

_____

_____

_____ 2. Fire departments should restrict the number of responses made to emergencies not involving fire.

_____

_____

_____ 3. Firefighters often find themselves in extraordinary situations that require them to exceed normal human limitations in order to bring an emergency to a safe conclusion.

_____

_____

_____ 4. Even after becoming certified, firefighters must continue their education through reading and studying assigned materials.

_____

_____

_____ 5. Fire police personnel assist regular police officers with traffic control, crowd control, and scene security during emergency operations.

_____

_____

_____ 6. A special rescue technician performs both topside and underwater rescues and recoveries.

_____

_____

_____ 7. An effective fire prevention program decreases the need for suppression activities and thereby reduces the cost and risk of extinguishing fire.

_____ 8. In departments that provide first response to EMS incidents, trained first-aid responders accompany regular firefighters on engines, trucks, or squads.

_____ 9. Many firefighters opt to learn new ideas, tactics, and methods throughout their careers, but training officially ends when a firefighter becomes certified.

_____ 10. Fire department members naturally cooperate effectively with one another because of the fraternity of their profession.

_____ 11. Firefighters should consider any instruction given on the fireground to be an order.

_____ 12. Personnel should never adjust procedures just to suit the situation.

_____ 13. Firefighters use SOPs primarily as a means to start the fire attack.

_______ 14. Fire personnel replace SOPs with size-up decisions based on professional judgment, evaluation, or command.

_______ 15. Firefighters should only perform fire attack SOPs while wearing complete protective clothing and SCBA.

_______ 16. Fire departments should limit the use of SOPs to the emergency scene.

_______ 17. All firefighters need *some* first-aid training.

_______ 18. Firefighters work closely with utility companies on many incidents that involve electricity, gas, and water utilities.

_______ 19. Most firefighter injuries cannot be prevented.

_______ 20. An assistance program is for fire personnel only.

_______ 21. Firefighters involved in operations that are extremely gruesome and horrific are required to participate in the CISD process.

_________ 22. If firefighters work more than one shift in horrific conditions, they should undergo a minor debriefing, called defusing, at the end of each shift.

____

____

_________ 23. Firefighters can safely stand on the tailboard of an apparatus while en route to an emergency.

____

____

_________ 24. Statistically, back injuries are the most expensive type of worker's compensation claims.

____

____

_________ 25. While attached, cheaters make tools unsafe to use, but luckily they do not permanently weaken tools.

____

____

_________ 26. Trainees with colds, headaches, or other physical discomforts during training drills should continue training unless they have a written excuse from a physician.

____

____

_________ 27. Fire departments must maintain records on all equipment used in training evolutions and use only equipment in excellent condition for training.

____

____

_________ 28. An IC should not restrict the efforts of fire personnel just because a rescue attempt puts a firefighter in danger.

____

____

_______ 29. A good personnel accountability system monitors individuals responding to the scene in both departmental apparatus and other vehicles.

____

____

Multiple Choice

C. Write the letter of the best answer on the blank before each statement.

___ 1. For what are all fire departments responsible?

a. Building inspection
b. Medical treatment and evacuation
c. Arson investigation
d. Life safety

___ 2. What factors determine the outcome of an emergency situation?

a. Extraordinary people, strategy, and safety
b. Knowledge, ability, and skill
c. Cooperation, calculated risk, and skill
d. Strategy, luck, and heroic gestures

___ 3. Why does the fire service need division of labor?

a. To prevent duplication of effort
b. To withhold specific responsibility
c. To make general assignments
d. To give everyone an assignment

___ 4. What personnel make up a company?

a. Fire chief, driver/operator, and two or more firefighters
b. Fire chief, driver/operator, and one or more firefighters
c. Company officer(s), driver/operator, and two or more firefighters
d. Company officer(s), driver/operator, and one or more firefighters

___ 5. Which company deploys hoselines for fire attack and exposure protection?

a. Truck (ladder) company
b. Rescue squad company
c. Engine company
d. Brush company

___ 6. Which company performs forcible entry, search and rescue, and ventilation?

a. Emergency medical unit
b. Hazardous materials company
c. Engine company
d. Truck (ladder) company

____ 7. What is a typical duty of a firefighter?
   a. Putting the safety of a victim above own safety
   b. Assuming responsibility for victims who cannot be saved
   c. Taking risks to preserve property
   d. Performing salvage operations

____ 8. What duties do information systems personnel perform?
   a. Receive emergency and nonemergency phone calls
   b. Maintain all fire department apparatus
   c. Manage electronic databases
   d. Dispatch units and provide a communication link

____ 9. Which NFPA standard addresses the duties of a hazardous materials technician?
   a. 472
   b. 1003
   c. 1021
   d. 1061

____ 10. What does a fire protection engineer do?
   a. Acts as a consultant to upper administration
   b. Informs the public about fire hazards
   c. Conducts technical and supervisory work in the fire prevention program
   d. Conducts investigations of a fire area to determine the origin of a fire

____ 11. Which of the following job descriptions best suits an EMT?
   a. Sustains life until more competent medical personnel arrive
   b. Provides basic life support for those whose lives are in danger
   c. Provides advanced life support
   d. Performs major medical procedures (such as amputation of a limb) needed to free a victim

____ 12. Which NFPA standard outlines the requirements for training officers and instructors?
   a. 1031
   b. 1035
   c. 1033
   d. 1041

____ 13. What is the most commonly accepted order of fireground priorities?
   a. Life safety, incident stabilization, and property conservation
   b. Incident stabilization, property conservation, and overhaul
   c. Life safety, property conservation, and overhaul
   d. Incident stabilization, overhaul, and property conservation

____ 14. What precaution should be taken at a medical response?
   a. Avoid performing mouth-to-mouth resuscitation.
   b. Wear rubber gloves and safety goggles.
   c. Remove all jewelry (e.g., watches and rings) that might trap or collect germs.
   d. Ignore safety precautions if necessary to save a life.

____ 15. In what situation would hospital personnel be called to the scene of an accident to assist a fire department?
   a. Mass casualty accident in which triage is necessary
   b. Automobile accident involving more than three vehicles
   c. Automobile accident involving hazardous materials
   d. Any incident involving the release of hazardous materials

____ 16. For what duties might law enforcement officers request assistance from fire service personnel?
   a. Collection of evidence at a crime scene
   b. Apprehension of suspects
   c. Emergency lighting at a crime scene
   d. Questioning possible arson suspects

____ 17. What factors motivate accident control within the fire fighting profession?
   a. Property conservation and economy
   b. Occupational safety and property conservation
   c. Economy and occupational safety
   d. Life safety and economy

____ 18. How do fire service officers instill good safety practices in their subordinates?
   a. Teach safety, and trust firefighters to follow the guidelines without a need for strict enforcement.
   b. Practice safety, and assume that others will follow the example.
   c. Recommend good safety practices.
   d. Teach, practice, and enforce safety.

____ 19. What basic physical conditioning should firefighters maintain?
   a. Short-distance sprinting, upper body strength, and overall muscle development
   b. Aerobic endurance, strength, and flexibility
   c. Flexibility, overall muscle development, and strength
   d. Strength, aerobic endurance, and short-distance sprinting

____ 20. Who is responsible for ensuring that measures are taken to limit the number of stress-related accidents and illnesses?

a. Firefighters
b. Fire departments
c. Safety officers
d. Fire chiefs

____ 21. In what areas does an EAP assist firefighters?

a. Borrowing money for personal expenses
b. Securing a loan for a new car
c. Finding and buying a new home
d. Any work-related or personal problem

____ 22. When should the CISD process start?

a. Before firefighters enter the scene
b. During rest periods at the scene
c. After firefighters leave the scene
d. At the firefighter's request

____ 23. When should firefighters *not* use handrails to dismount an apparatus?

a. When an apparatus is parked near a power station
b. When an aerial device is extended close to electrical wires
c. During an electrical storm
d. During a fire attack on a large electrical fire

____ 24. Who is responsible for young visitors to the fire station?

a. Parents
b. Chaperones
c. Fire department personnel
d. Whoever signed the field trip approval slip

____ 25. What usually causes a slip, trip, or fall?

a. Horseplay
b. Poor footing
c. Waxed floors
d. Inadequate lighting

____ 26. Which NFPA standard regulates the design, construction, usage, maintenance, inspection, and repair of fire fighting tools and equipment?

a. 1500
b. 1021
c. 1521
d. 1061

____ 27. How does PPE effect tool and equipment use?

a. Reduces the dangers of using tool extensions
b. Replaces the need for standardized safety guards
c. Compensates for poor tool design
d. Provides firefighters with protection against hazards

____ 28. Which of the following guidelines must be included in power tool safety?
   a. Keep manufacturer instructions in a locked file cabinet to prevent them from being lost.
   b. Post one easy-to-follow instruction sheet for all air driven tools and one for all electric tools.
   c. Keep accurate records of repairs made to all tools.
   d. Follow the instructions of an experienced power tool user.

____ 29. What rules should be followed when using tools that are *not* marked double insulated?
   a. Remove the third prong to avoid unpredictable electrical shorts.
   b. Bypass the ground plug when necessary.
   c. Never use electrical tools that are not double insulated.
   d. Always connect the ground plug while using the tool.

____ 30. According to NFPA 1500, how often must personnel who engage in structural fire fighting participate in training?
   a. Monthly
   b. Biannually
   c. Quarterly
   d. Annually

____ 31. When are firefighters expected to participate in special training?
   a. When they are unsuccessful in a fire attack
   b. When new equipment is introduced
   c. When they are involved in an accident
   d. When a firefighter dies during an emergency incident

____ 32. Why do firefighters most often engage in horseplay and disrupt training classes?
   a. The trainees want to impress the instructor with their humor.
   b. The trainees are not suitable candidates for fire fighting.
   c. The trainees cannot see the demonstrations.
   d. The trainees have social problems that prevent them from paying attention.

____ 33. When do fire departments expect firefighters to take excessive risks to their own safety?
   a. During a rescue attempt that involves three or more people
   b. While en route to an emergency
   c. When an incident is understaffed
   d. Never

____ 34. How should firefighters deal with relatives and friends of victims?
  a. Remove them from the cordoned area.
  b. Leave them alone to console each other privately.
  c. Restrain them from getting too close.
  d. Assure them that everything will be okay.

____ 35. What factor determines the specific distance from the emergency scene or area that should be cordoned off?
  a. The space provided by apparatus placement
  b. The amount of space secured by police before the crowd got too large
  c. The number of bystanders getting in the way
  d. The area needed by personnel to work

____ 36. How does a tag system aid in tracking fire service personnel?
  a. Firefighters leave personal identification tags with designated personnel while inside the fireground perimeter.
  b. Command post officers provide tags to personnel as they enter the fireground and retrieve the tags as firefighters leave.
  c. Firefighters wear identification tags only while on duty, whether in the fireground perimeter or not.
  d. Firefighters use tags to identify locations they have searched during rescue and recovery.

Identify

D. Identify the following abbreviations associated with orientation and safety. Write the correct interpretation before each.

________________________ 1. NFPA

________________________ 2. ARFF

________________________ 3. NBC

________________________ 4. SCUBA

________________________ 5. EMS

________________________ 6. EMT

_______________ 7. BLS

_______________ 8. ALS

_______________ 9. SOP

_______________ 10. SCBA

_______________ 11. IV

_______________ 12. EPA

_______________ 13. EAP

_______________ 14. CISD

_______________ 15. PPE

FIREFIGHTER II

Matching

A. Match to their definitions terms associated with IMS components. Write the appropriate letters on the blanks.

____ 1. Responsible for managing all operations that directly affect the primary mission

____ 2. Responsible for the collection, evaluation, dissemination, and use of information concerning the development of the incident

____ 3. Responsible for providing the facilities, services, and materials necessary to support the incident

____ 4. Responsible for all incident activities, including the development and implementation of a strategic plan

a. Planning
b. Command
c. Finance
d. Logistics
e. Operations

B. Match to their definitions terms associated with IMS. Write the appropriate letters on the blanks.

____ 1. The function of directing, ordering, and controlling resources by virtue of explicit legal, agency, or delegated authority

____ 2. A geographic designation assigning responsibility for all operations within a defined area

____ 3. A geographic *or* functional assignment equivalent to a division or a group

____ 4. The person in command of a division, group, or sector

____ 5. The written or unwritten plan for managing the emergency

____ 6. The officer responsible for everything that takes place on the emergency scene

____ 7. All personnel and major pieces of apparatus on the scene or en route to the scene

a. Group
b. Division
c. Resources
d. Command
e. Sector
f. IC
g. IAP
h. Supervisor

True/False

C. Write *True* or *False* before each of the following statements. Correct those statements that are false.

________ 1. Fire departments should apply IMS to large-scale, complex incidents only.

____

____

________ 2. The IC has the authority to call resources to an incident as well as release resources from it.

____

____

________ 3. Personnel should subdivide the IMS Operations component into no more than three branches when necessary.

____

____

________ 4. Departments should activate the IMS Finance component at all incidents, whether routine or large-scale.

____

____

________ 5. Firefighters must immediately and without question follow lawful commands given by those in authority.

____

____

________ 6. Personnel should formulate a written plan for every incident.

____

____

_______ 7. Fire service personnel must track the status of all resources to determine which ones are available for assignment when needed.

_______ 8. The Operations Officer gathers resources and organizes information to ensure that orders can be carried out promptly, safely, and efficiently.

_______ 9. Only company officers and section officers can deviate from the IAP during an incident.

_______ 10. The first arriving fire service personnel usually transfers command to the next arriving person with a higher level of expertise or authority.

_______ 11. Personnel can only transfer command to a qualified fire service official, either on the scene or en route to the scene.

_______ 12. A senior member of the fire service arriving at the scene may choose to leave a subordinate in command of an incident.

_______ 13. Although IMS provides a means for tracking personnel and equipment at the scene of an incident, this function is not usually necessary.

_______ 14. The IC does not release any resources until an incident has been successfully resolved.

____

____

Multiple Choice

D. Write the letter of the best answer on the blank before each statement.

____ 1. What components are necessary to provide a basis for clear communication and effective operations?
- a. Current telecommunication devices and modern computer equipment
- b. Singular communications and qualified communications engineers
- c. Individualized action plans and diversified command structure
- d. Common terminology and comprehensive resource management

____ 2. What position falls within the scope of the Command Staff?
- a. Operations Officer
- b. Safety Officer
- c. Logistics Officer
- d. Technical specialists

____ 3. Into what two areas does IMS Logistics break down?
- a. Resource Unit and Situation Status Unit
- b. Demobilization Unit and technical specialists
- c. Support branch and service branch
- d. Rescue branch and recovery unit

____ 4. How are divisions usually assigned at an emergency scene?
- a. Counterclockwise around an outdoor incident
- b. Via right or left at a fire involving two adjacent buildings
- c. As north, south, east, or west in a one-floor building
- d. By floor in a multilevel building

____ 5. To whom should all groups or functional sectors operating within a specific geographic area report?
- a. Liaison Officer
- b. Safety Officer
- c. Division supervisor
- d. Operations Officer

____ 6. Which of the following statements best describes the term "groups?"
   a. When the assigned function has been completed, they are available for reassignment.
   b. Until the incident is resolved, they are not available for reassignment.
   c. Throughout their shift, they remain together, but are available for reassignment for following shifts.
   d. Unless reassigned by an officer, they serve together during every incident.

____ 7. Who is ultimately responsible for everything that takes place at an emergency scene?
   a. IC
   b. Safety Officer
   c. Operations Officer
   d. Planning Officer

____ 8. When are resources considered available?
   a. Upon leaving the fire station if not already committed
   b. As soon as they arrive at the scene
   c. After checking in and if not already assigned
   d. Only when approved for assignment by the IC

____ 9. Who initiates IMS during an emergency?
   a. The dispatcher/telecommunicator taking the emergency call
   b. The first company officer arriving at the scene
   c. The first fire service person arriving at the scene
   d. The highest ranking officer en route to the scene

____ 10. According to the IAP, which of the following actions should be the highest priority at an emergency scene?
   a. Rescuing or evacuating endangered occupants
   b. Eliminating the hazard
   c. Conducting loss control
   d. Ensuring personnel safety and survival

____ 11. How many ICs should be used at a typical emergency?
   a. One per incident
   b. Two per incident
   c. One per IMS component
   d. Two per IMS component

____ 12. During a multijurisdictional incident with a unified command, how does the IC position change?
   a. Each jurisdiction sets up their own command post.
   b. The chain of command becomes less clearly defined.
   c. More than one person issues orders through the chain of command.
   d. More than one person serves as IC.

____ 13. What information should be included in a situation status report?
  a. A description of what happened, what has been done, and what resources are on the scene or en route
  b. An account of how the emergency may have started, the names of the departments on hand, and the number of victims recovered
  c. A detailed narrative about the source of the emergency, the various types of extinguishing agents or rescue tactics being used, and the number of firefighters and medical personnel already at the scene
  d. A count of victims who survived or perished, an estimate of cost damage, and an approximate length of time the incident will last

____ 14. How should the person relinquishing command render a situation status report?
  a. Write a synopsis of what has occurred and request that the new commander read the incident evaluation aloud.
  b. State the incident evaluation and ask the new commander to repeat the evaluation to ensure the information is correct.
  c. Narrate the incident evaluation to the new commander three times.
  d. Allow the new commander to evaluate the situation and fill in pertinent information as necessary.

____ 15. Why should the former IC announce a command change?
  a. To prevent incident personnel from reporting to the wrong commander and thereby disrupting the chain of command
  b. To avoid any possible confusion caused by others hearing a different voice acknowledging messages and issuing orders
  c. To advise workers at the scene that a new action plan will probably be implemented
  d. To indicate that an officer of higher authority has arrived on the scene

____ 16. Following the chain of command and using correct radio protocols, how should personnel address others on the radio?
  a. By name
  b. By rank
  c. By job title
  d. By message only

____ 17. What should the size and complexity of the IMS reflect?
  a. The size of the department in command
  b. The number of personnel available for assignment
  c. The magnitude of the situation
  d. The level of expertise of the IC

____ 18. What procedures are included in an IAP tracking and accountability system?

a. Checking in at the scene, identifying the location of personnel on the scene, and releasing units no longer needed

b. Setting up a command center at the fire station, routing apparatus to the scene, and mapping the progress of the personnel on location

c. Assigning personnel to units prior to an emergency, identifying apparatus en route to an emergency, and locating personnel for shift changes

d. Mapping the progress of personnel on the scene, locating personnel for shift changes, and releasing personnel no longer needed

Identify

E. Identify the following abbreviations associated with IMS. Write the correct interpretation before each.

______________________________ 1. IMS

______________________________ 2. IC

______________________________ 3. IAP

______________________________ 4. CP

Chapter 2
Fire Behavior

Chapter 2 Fire Behavior

FIREFIGHTER I

Matching

A. Match to their definitions terms associated with fire behavior. Write the appropriate letters on the blanks.

____ 1. The science of fire and the factors that effect its ignition, growth, and spread

____ 2. A chemical reaction that requires fuel, oxygen, and heat to occur

____ 3. The study of the physical world, including chemistry, physics, and laws related to matter and energy

____ 4. The material or substance being oxidized or burned in the combustion process

____ 5. The chemical decomposition of a substance through the action of heat

____ 6. The surface area of the fuel in proportion to the mass

____ 7. The transformation of a liquid to its gaseous state

____ 8. The total amount of fuel in a specific location multiplied by the heat of combustion of materials

____ 9. A series of reactions that occur in sequence with the results of each individual reaction being added to the rest

____ 10. The tendency of gases to form into layers according to temperature

a. Fire

b. Pyrolysis

c. Upper flammable limit

d. Physical science

e. Vaporization

f. Thermal layering

g. Fire behavior

h. Surface-to-mass ratio

i. Chain reaction

j. Fuel

k. Fuel load or fire load

B. **Match to their definitions terms associated with physical science. Write the appropriate letters on the blanks.**

____ 1. The measurement of the gravitational attraction on a specific mass

____ 2. The capacity to perform work

____ 3. The transformation of energy from one form to another

____ 4. The energy possessed by a moving object

____ 5. An amount of energy delivered over a given period of time

____ 6. The energy transferred from one body to another when the temperatures of the bodies are different

____ 7. An indicator of heat and the measure of the warmth or coldness of an object based on some standard

____ 8. The amount of heat required to raise the temperature of 1 gram of water 1 degree Celsius

____ 9. The amount of heat required to raise the temperature of 1 pound of water 1 degree Fahrenheit

____ 10. Anything that occupies space and has mass

a. Work
b. Calorie
c. Weight
d. Btu
e. Power
f. Energy
g. Matter
h. Potential energy
i. Kinetic energy
j. Temperature
k. Heat

C. Match the International System of measurement units to their equivalent Customary System units. Write the appropriate letters on the blanks.

Customary System

____ 1. Pound (lb)

____ 2. Foot-pound (ft lb)

____ 3. Horsepower (hp)

____ 4. Fahrenheit (F)

____ 5. Calorie (Cal) or British thermal unit (Btu)

____ 6. Btu per second (Btu/s)

____ 7. Pound per cubic foot (lb/ft^3)

International System of Units

a. Celsius (C)

b. Joule (J)

c. Kilogram per cubic meter (kg/m^3)

d. Watts (W)

e. Gigawatts (gW)

f. Newtons (N)

g. Kilowatts (kW)

True/False

D. Write *True* or *False* before each of the following statements. Correct those statements that are false.

________ 1. Firefighters should attack structural and wildland fires the same way.

____

____

________ 2. The scientific community, as well as most nations other than the United States, use the International System of Units.

____

____

________ 3. The English or Customary System logically bases all measurements on powers of 10.

____

____

_______ 4. A vacuum contains no medium for the point-to-point contact needed to conduct heat.

____

____

_______ 5. A good insulator sustains point-to-point heat transfer.

____

____

_______ 6. Radiation causes most exposure fires.

____

____

_______ 7. Temperature, alone, determines at what point water will boil.

____

____

_______ 8. Liquids with a specific gravity less than 1 are lighter than water, while those with a specific gravity greater than 1 are heavier than water.

____

____

_______ 9. Researchers can calculate the heat release rate of materials by using instruments that determine mass loss and temperature gain when a fuel is burned.

____

____

_______ 10. Oxidation absorbs energy or is endothermic.

____

____

_______ 11. The slowness of the chemical reaction between oxygen and iron prevents the generation of heat.

____

____

_________ 12. The fire triangle more accurately depicts the components of a fire than does the fire tetrahedron.

____

____

_________ 13. Only oxygen causes oxidation.

____

____

_________ 14. When placed in an oxygen-enriched atmosphere, even some fire-resistant materials like Nomex® ignite and burn vigorously.

____

____

_________ 15. Normally, fuels must be in a gaseous state to burn.

____

____

_________ 16. The volatility or ease with which a liquid gives off vapor influences its ignitability.

____

____

_________ 17. When contained, the specific volume of a liquid has a relatively high surface-to-volume ratio.

____

____

_________ 18. For standardization, researchers normally express any heat of combustion in terms of the heat of gasoline combustion.

____

____

_________ 19. Chemical heat, like the heat generated when a match burns, is the most common source of heat in combustion reactions.

____

____

_______ 20. Ignition describes the period when the three elements of the fire triangle come together and combustion begins.

____

____

_______ 21. Flashover occurs when temperatures reach 1,200°F *(649°C)*.

____

____

_______ 22. Occupants who have not escaped from a compartment before flashover occurs are not likely to survive.

____

____

_______ 23. Personal protective gear safeguards firefighters from the extreme danger of a compartment flashover.

____

____

_______ 24. During decay, the fire becomes fuel controlled, the amount of fire diminishes, and the temperatures within the compartment begin to decline.

____

____

_______ 25. Firefighters have no way to determine the fire growth potential for a building or space.

____

____

_______ 26. Flameover involves only the fire gases, not the surfaces, of the other fuel packages within a compartment.

____

____

_______ 27. Improper ventilation can allow air to mix with hot gases and cause an explosive ignition called backdraft.

____

____

_______ 28. The heat energy from fires causes more deaths than smoke inhalation.

____

____

_______ 29. Smoke contains toxic substances, most commonly CO, that can be deadly to firefighters not wearing SCBA.

____

____

_______ 30. Water effectively extinguishes ignitable liquids with a specific gravity of less than 1.

____

____

_______ 31. Polar solvents dissolve in water.

____

____

_______ 32. Hydrocarbon gases have vapor densities greater than 1, which allow them to rise and dissipate when released.

____

____

_______ 33. Each fire class requires different extinguishing agents.

____

____

_______ 34. CO_2 flooding provides adequate cooling for Class A fires.

____

____

_______ 35. Oxygen-exclusion extinguishment methods work best on fires involving flammable and combustible liquids and gases.

____

____

_______ 36. No single agent controls fires in all combustible metals.

____

____

_______ 37. Firefighters should immediately begin fighting Class D fires with a blanket of dry chemicals.

____

____

Multiple Choice

E. Write the letter of the best answer on the blank before each statement.

____ 1. What is the length base unit for SI?
   a. Foot
   b. Meter
   c. Mole
   d. Candela

____ 2. Energy exists in which two states?
   a. Kinetic and potential
   b. Chemical and mechanical
   c. Electrical and nuclear
   d. Heat and light

____ 3. What type of energy transfers between two bodies of differing temperatures such as the sun and the earth?
   a. Light
   b. Nuclear
   c. Heat
   d. Chemical

____ 4. What factors determine how much power is needed to complete a task?
   a. Amount of energy and period of time
   b. Period of time and amount of work
   c. Amount of work and distance
   d. Distance and amount of energy

____ 5. What is the mechanical equivalent of heat?
   a. 1 calorie equals 1,055 Btu
   b. 1 calorie equals 4.187 Btu
   c. 1 calorie equals 1,055 J
   d. 1 calorie equals 4.187 J

____ 6. How does temperature affect the rate of heat transfer?
   a. The greater the heat of the two bodies, the greater the transfer rate
   b. The greater the difference in temperature between the two bodies, the greater the transfer rate
   c. The lower the heat of the two bodies, the greater the transfer rate
   d. The lower the difference in temperature between the two bodies, the greater the transfer rate

____ 7. What mechanisms transfer heat from one body to another?
   a. Confection, reflection, and refraction
   b. Reflection, conduction, and radiation
   c. Refraction, reflection, and convection
   d. Conduction, convection, and radiation

____ 8. As depicted below, what type of heat transfer moves from point to point?
   a. Conduction
   b. Convection
   c. Confection
   d. Radiation

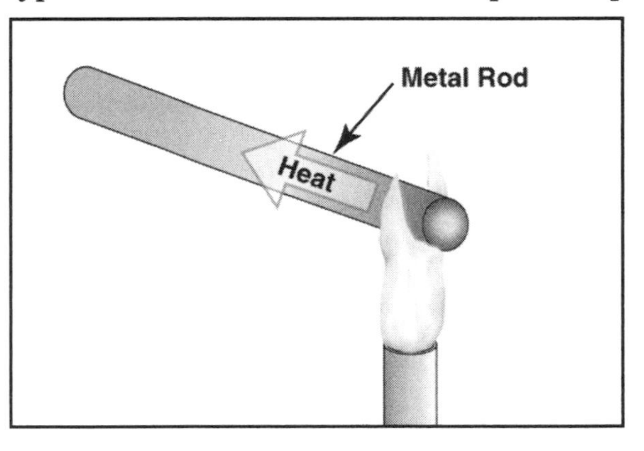

____ 9. As depicted below, what type of heat transfer occurs when a fluid or gas flows from one place to another?
   a. Conduction
   b. Convection
   c. Radiation
   d. Reflection

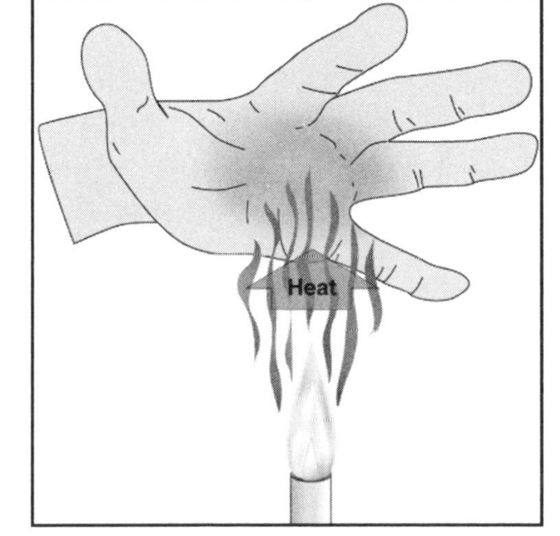

____ 10. As depicted below, what type of heat transfer occurs without an intervening medium?

a. Conduction
b. Convection
c. Radiation
d. Refraction

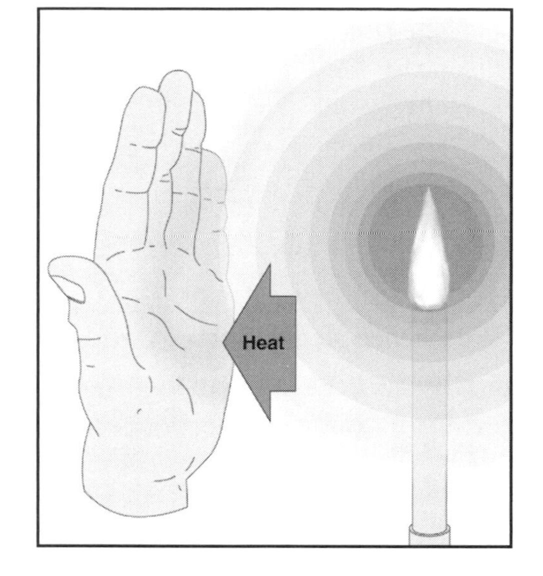

____ 11. What disrupts the transmission of radiated energy?

a. Vacuums
b. Substantial air spaces
c. Reflective materials
d. Refractive materials

____ 12. What factors affect the physical state of matter?

a. Temperature and pressure
b. Substance and form
c. Surface and temperature
d. Pressure and form

____ 13. What term best describes the ratio of the mass of a given volume of liquid compared with the mass of an equal volume of water?

a. Vapor density
b. Matter density
c. Liquid mass
d. Specific gravity

____ 14. What basic principle does the Law of Conservation of Mass teach?

a. Mass and energy are neither created nor destroyed.
b. Mass can be created but not destroyed.
c. Energy can be created but not destroyed.
d. Mass and energy can be both created and destroyed.

____ 15. What process transforms matter from one state to another or produces a new substance?

a. Heat reaction
b. Physical reaction
c. Chemical reaction
d. Chain reaction

____ 16. What reaction gives off energy as it occurs?

a. Endothermic
b. Rusting
c. Refraction
d. Confection

____ 17. What reaction is produced when iron and oxygen combine?

a. Oxidation
b. Endothermic rusting
c. Chemical attachment
d. Heat transfer

____ 18. Explosions are the result of what kind of reaction?

a. A slow reaction of a fuel and an oxidizer
b. A slow reaction of a fuel and heat
c. A rapid reaction of a fuel and an oxidizer
d. A rapid reaction of a fuel and heat

____ 19. What components must come together for combustion to occur?

a. Oxidizing agent, fuel, and carbon dioxide
b. Oxidizing agent, fuel, and a self-sustained chemical reaction
c. Oxidizing agent, fuel, heat, and carbon dioxide
d. Oxidizing agent, fuel, heat, and a self-sustained chemical reaction

____ 20. What effect does an oxygen-enriched atmosphere have on fire?

a. Some petroleum-based materials autoignite.
b. Most materials exhibit the same burning characteristics.
c. Many materials that burn at normal oxygen levels do not burn at elevated levels.
d. All of the above effects occur.

____ 21. What two factors influence the combustion process?

a. Distribution of the fuel and reducing agent
b. Physical state and distribution of the fuel
c. Volume of the fuel and reducing agent
d. Mass and physical state of the fuel

____ 22. Which of the following wood products has the highest surface-to-mass ratio?

a. Logs
b. Sawdust
c. Boards
d. Tree stumps

____ 23. What happens to fuel as its surface area increases?

a. Less material becomes exposed to the heat.
b. Materials generate less burnable gases due to pyrolysis.
c. Fuel particles become smaller, and ignitability increases.
d. Less fuel becomes available, and ignitability decreases.

____ 24. How does positioning affect the way a solid fuel burns?

a. Fire spreads more rapidly in vertically positioned solid fuels.
b. Fire spreads more rapidly in horizontally positioned solid fuels.
c. Positioning does not affect the way solid fuels burn.
d. Positioning only affects solid fuels in a vacuum.

____ 25. What fuel type already exists in the natural state required for ignition?
a. Solid
b. Liquid
c. Gas
d. Compressed

____ 26. Gaseous fuels must mix with the proper ratio of oxidizer for combustion to occur. What is that ratio?
a. Below LFL
b. Above UFL
c. Between LFL and UFL
d. Below LFL or above LFL

____ 27. In general, how does an increase in temperature affect the flammable range of a material?
a. Narrows it
b. Broadens it
c. Levels it
d. Streamlines it

____ 28. Which of the following items would *not* be considered a fuel package?
a. A mattress and box spring
b. A foam-padded upholstered chair
c. A computer and office furniture
d. A stack of bricks

____ 29. What is the total amount (mass) of fuel in a compartment multiplied by the heat of the combustion of the materials?
a. Fuel package
b. Fuel load
c. Flammable range
d. Heat of combustion

____ 30. How is combustion influenced when heat comes into contact with a fuel?
a. Reduces the production of ignitable vapors
b. Prevents pyrolysis or vaporization of solid and liquid fuels
c. Provides energy for ignition
d. Inhibits the continuous production and ignition of fuel vapors

____ 31. What events must occur for spontaneous combustion to happen?
a. The rate of heat production must raise the temperature to within 10 degrees of its ignition temperature.
b. The air supply to the material being heated must be undetectable.
c. The materials surrounding the fuel must prevent the heat from dissipating.
d. The ignitable vapors must come into contact with an ignition source.

____ 32. Which of the following materials is most likely to spontaneously heat?
a. Linseed oil rags
b. Fertilizer
c. Hay
d. Iron metal powder

____ 33. What type of heat energy is generated by friction and compression?
a. Electrical
b. Mechanical
c. Nuclear
d. Physical

____ 34. What characterizes slow oxidation reactions?
a. They do not produce heat fast enough to reach ignition.
b. They never generate sufficient heat to become self-sustained.
c. They are exothermic reactions.
d. They have all of the above characteristics.

____ 35. What factors control the growth and development of a compartment fire?
a. Size and duration
b. Duration and fuel
c. Fuel and ventilation
d. Ventilation and size

____ 36. Which fuel packages entrain the least air and have the highest plume temperatures?
a. Fuel packages in corners
b. Fuel packages against walls
c. Fuel packages in the middle of a room
d. Fuel packages near windows or doors

____ 37. Which of the following scenarios best describes flashover?
a. The transition between the growth and full development of a fire
b. A fully developed fire
c. The ignition of a fuel package by radiant heat
d. The decay stage of a compartment fire

____ 38. What is the HRR?
a. The amount of light being released over time
b. The amount of heat being released over time
c. The amount of fuel being released over time
d. The amount of gas being released over time

____ 39. Which of the following materials has the highest HRR?
a. Wastebasket full of milk cartons
b. Cotton mattresses
c. Wooden pallets
d. Metal chairs with cotton padding

____ 40. How may heat generated in a compartment fire transmit from the initial fuel package to other fuels?
a. Refraction
b. Reflection
c. Confection
d. Radiation

____ 41. How should firefighters deal with thermal layering?
a. Apply water to the upper level of the layer.
b. Avoid disrupting the thermal balance.
c. Ventilate the structure horizontally.
d. Avoid working in the lower level of the layer.

____ 42. What condition signals a possible backdraft?
a. Billowing black smoke exiting large openings
b. Large flames
c. Smoke-stained windows
d. Smoke leaving a building in a continuous motion

____ 43. What common narcotic gases are found in smoke?
a. CO, HCL, and N_2O
b. N_2O, HCL, and CO_2
c. CO_2, CLN, and HCN
d. HCN, CO, and CO_2

____ 44. What does flooding an area with an inert gas do to a fire?
a. Reduces the temperature
b. Eliminates the available fuel
c. Disrupts the combustion process
d. Creates a fire-resistant barrier

____ 45. How does water combat fire?
a. Reduces the temperature
b. Eliminates the available fuel
c. Disrupts the combustion process
d. Creates a fire-resistant barrier

Identify

F. Identify the following abbreviations associated with fire behavior. Write the correct interpretation before each.

______________________________ 1. SI

______________________________ 2. NIST

______________________________ 3. HRR

______________________________ 4. LFL

_______________ 5. UFL

_______________ 6. CO

_______________ 7. HCN

_______________ 8. CO_2

_______________ 9. MSDS

_______________ 10. NAERG

G. Identify the following stages of fire development. Write the correct fire stage below each picture.

_______________ 1.

_______________ 2.

• Superheated vapors ignite
• Flame front rolls across ceiling

________________________ 3.

• Introduction of oxygen causes fire of explosive force.

________________________ 4.

H. Identify the following methods of fire extinguishment. Write the correct letter next to the description given.

_____ 1. Exclusion of oxygen

_____ 2. Reduction of temperature

_____ 3. Inhibition of chain reaction

_____ 4. Removal of fuel

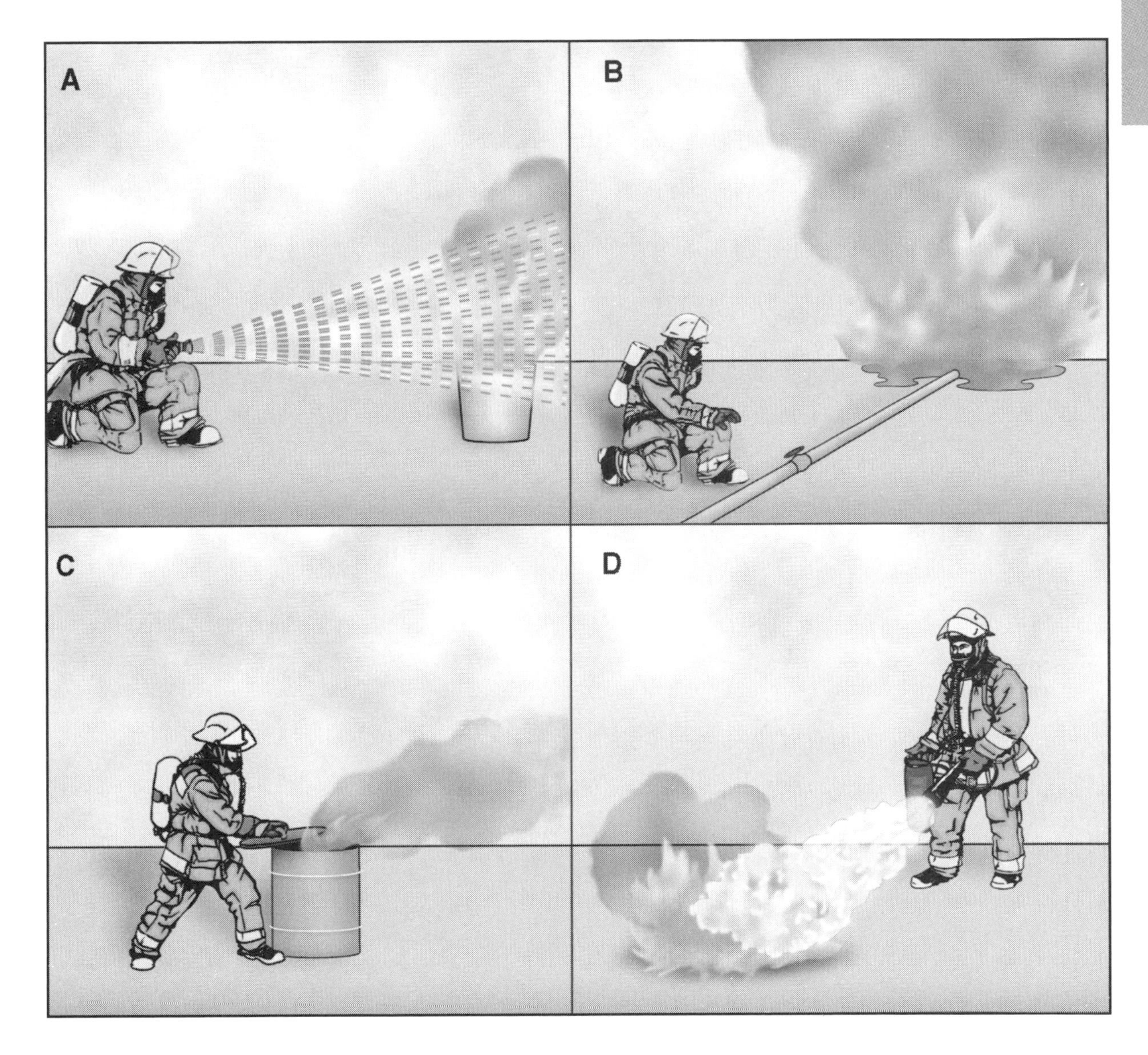

I. Identify the following fuels. Write the correct fuel class below each picture.

____________________ 1.

____________________ 2.

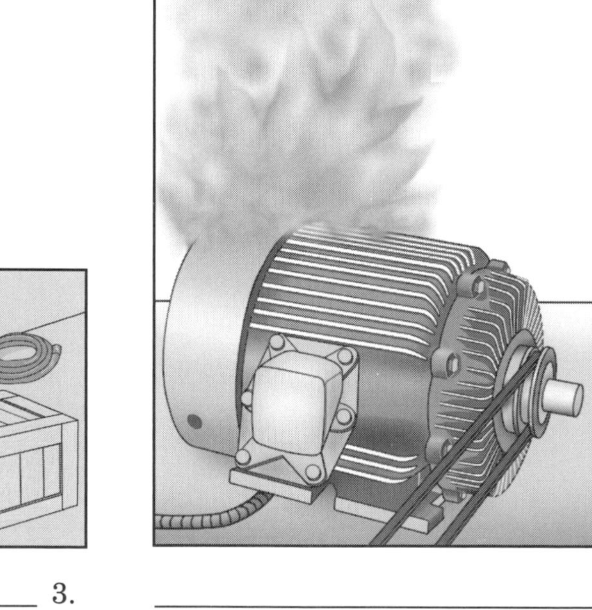

____________________ 3.

____________________ 4.

Chapter 3
Building Construction

Chapter 3 Building Construction

FIREFIGHTER I

Matching

A. Match to their definitions terms associated with building construction. Write the appropriate letters on the blanks.

____ 1. Components needed to provide a separating fire wall that meets the requirements of a specified fire-resistance rating

____ 2. Inorganic product from which plaster and plasterboard are made

____ 3. The maximum heat that can be produced if all the combustible materials in a given area burn

____ 4. The presence of large amounts of combustible materials in an area of a building

____ 5. Vertical and horizontal stresses that tend to pull things apart

a. Gypsum
b. Fire loading
c. Tension
d. Fire wall assemblies
e. Fire load
f. Compression

True/False

B. Write *True* or *False* before each of the following statements. Correct those statements that are false.

________ 1. Only safety officers must have a basic knowledge of the principles of building construction.

____

____

________ 2. Partitions and floors constructed with noncombustible or limited combustible materials tend to retard the spread of fire through a building.

____

____

_______ 3. Type II construction contains no materials without fire-resistance ratings, such as untreated wood.

____

____

_______ 4. Concealed spaces are primary fire concerns specific to ordinary construction.

____

____

_______ 5. Type IV construction has virtually no concealed spaces.

____

____

_______ 6. Fire coming from the doors and windows of a wood-frame structure is a good sign because it means that the building is adequately ventilated.

____

____

_______ 7. Fire retardants effectively reduce the spread of fire.

____

____

_______ 8. Water used during extinguishing operations has a substantial negative effect on the structural strength of wood construction materials.

____

____

_______ 9. Bolts or other connections holding cast iron to a building can fail during a fire and cause sections of the metal to come crashing down.

____

____

_______ 10. Elongating steel can actually push out load-bearing walls and cause a collapse.

____

____

_______ 11. Reinforced concrete withstands fire exceptionally well.

_______ 12. Gypsum breaks down easily under fire conditions.

_______ 13. To make wood shake shingles safe, homeowners must treat them with fire retardant.

_______ 14. Large, open spaces in buildings contribute to the spread of fire.

_______ 15. Only trained safety personnel should be allowed to operate apparatus in a collapse zone.

_______ 16. Lightweight trusses that have been treated with fire retardants remain safe during a fire.

_______ 17. In any structure containing trusses, if one member fails, the entire truss will likely fail.

_______ 18. A building under construction or renovation is at higher risk for fire than other structures.

Multiple Choice

C. Write the letter of the best answer on the blank before each statement.

____ 1. How are construction classifications determined?
   a. Fire resistance of construction materials and hours of fire protection the structural components give
   b. Fire resistance of construction materials and distance of structure from a fire department
   c. Hours of protection the structural components give and distance of structure from a fire department
   d. Distance of structure from a fire department and fire code rating of structure

____ 2. Which of the following design features does *not* compromise the ability of fire-resistive construction to confine fire to a certain area?
   a. Openings made in partitions
   b. Improperly designed heating and air-conditioning systems
   c. Dampered heating and air-conditioning systems
   d. Compartmentation created by partitions and floors

____ 3. Which of the following fire hazards should *not* be a concern for fire protection in a Type II construction?
   a. The contents of the building may be highly combustible.
   b. Heat buildup can cause structural supports to fail.
   c. Flat, built-up roofs, which contain felt, insulation, and tar, can collapse.
   d. The interior construction may contain a large amount of untreated wood.

____ 4. What fire hazards do flat, built-up roofs containing felt, insulation, and roofing tar create?
   a. Fire extension to the roof can cause the entire roof to fail.
   b. Fire victims often flee to the roof and become trapped during a fire.
   c. Firefighters cannot reach the roof to perform proper ventilation.
   d. None are created; these roofs are the safest type for fire conditions.

____ 5. What materials completely or partially make up the interior structural members of Type III construction?
   a. Noncombustible
   b. Wood
   c. Steel
   d. Heavy timber

____ 6. What can be used to stop the spread of fire in concealed spaces?

a. Drywall
b. Gypsum
c. Plaster
d. Fire-stops

____ 7. What structures are most commonly Type V constructions?

a. Churches
b. Factories
c. Single-family residences
d. Barns

____ 8. What type of wall supports two adjacent structures?

a. Nonload-bearing wall
b. Party wall
c. Partition wall
d. Fire wall

____ 9. Which wood type takes the longest to burn?

a. Cured wood
b. Dried wood
c. Green wood
d. Treated wood

____ 10. How do composite building components, such as particleboard, fiberboard, and paneling, hold up during a fire?

a. They are not highly combustible.
b. They can produce significant toxic gases.
c. They slowly deteriorate under fire conditions.
d. They retard the spread of fire.

____ 11. What are the freestanding fire walls, commonly found on large churches and shopping centers, called?

a. Partition walls
b. Cantilever walls
c. Veneer walls
d. Party walls

____ 12. How is masonry affected by fire and exposure to high temperatures?

a. Bricks lose integrity and show serious deterioration.
b. Stones show no signs of deterioration.
c. Blocks lose strength and basic structural stability.
d. Masonry in general holds up well in these conditions.

____ 13. What does the rapid cooling effect of water do to masonry?

a. Causes spalling or cracking
b. Preserves structural integrity
c. Prevents fire damage
d. Weakens bricks, stones, or blocks

____ 14. What is the primary material used for structural support in modern building construction?

a. Steel
b. Reinforced concrete
c. Heavy timber
d. Fiberglass

____ 15. Which material elongates when heated?
a. Steel
b. Reinforced concrete
c. Gypsum
d. Fiberglass

____ 16. What factor determines the temperature at which specific steel members will fail?
a. Age of construction
b. Composition of the steel
c. Location of the member
d. Speed of the fire spread

____ 17. How does water affect steel members in fire conditions?
a. Shatters the member
b. Reduces the risk of failure
c. Increases the risk of structural collapse
d. Produces no effect

____ 18. What do cracks and spalls in reinforced concrete indicate to firefighters?
a. The construction has not sustained significant damage.
b. The structural strength of the building remains at optimal levels.
c. The concrete was incorrectly attached to the steel.
d. The concrete and steel bond has weakened.

____ 19. What is the most typical characteristic of conventional glass?
a. Serves as a barrier to fire extension
b. Creates an effective thermal protection barrier
c. Cracks and shatters when struck by a fire stream
d. Becomes a significant fuel source for fire

____ 20. What is a major fire hazard in commercial and storage facilities?
a. Faulty electrical wiring
b. Heavy fire loading
c. Defective sprinkler systems
d. Outdated fire extinguishers

____ 21. What is the most effective defense against fire hazards in storage facilities?
a. Proper code inspection and enforcement
b. Updated fire extinguishers and alarm systems
c. Well-maintained sprinkler systems
d. Correctly installed electrical wiring

____ 22. What two elements contribute to fire spread and smoke production and have been identified as major factors in the loss of lives in fires?
  a. Combustible furnishings and finishes
  b. Gypsum and plasterboard
  c. Fiberglass and reinforced glass
  d. Wood construction and particleboard

____ 23. How should firefighters slow the spread of fire in buildings with large, open spaces?
  a. Horizontal ventilation
  b. Vertical ventilation
  c. Horizontal ventilation followed by vertical ventilation
  d. No ventilation

____ 24. Which of the following buildings would *least* likely suffer structural failure due to the effects of fire?
  a. Truss construction grocery store
  b. Heavy timber church
  c. Lightweight construction single-family house
  d. Old wood construction warehouse

____ 25. What fire fighting operations increase the risk of building collapse?
  a. Putting water onto the structure
  b. Performing vertical ventilation between structural supports
  c. Standing on the roof of the structure
  d. Extinguishing the lower levels before the upper levels of a multilevel structure

____ 26. What dimensions should a collapse zone have?
  a. Equal to the height of the building
  b. Equal to one and a half times the height of the building
  c. Equal to two times the height of the building
  d. Equal to the safety officer's determination of a safe distance

____ 27. In what time frame do lightweight metal and wood trusses usually fail?
  a. 5 to 10 minutes
  b. 10 to 15 minutes
  c. 15 to 20 minutes
  d. 20 to 25 minutes

____ 28. What type of building is subject to faster-than-normal fire spread?
  a. Fully occupied office building
  b. Newly renovated family residence
  c. Modern apartment building
  d. Abandoned warehouse

____ 29. What action should firefighters take when arriving at a fire in a building with truss roof construction that has been on fire for 5 minutes?
   a. Enter the building and start the attack.
   b. Climb onto the roof and begin ventilation.
   c. Set up a collapse zone.
   d. Call for reinforcements before entering the building.

____ 30. What is the critical temperature for steel truss failure?
   a. 500°F (*260°C*)
   b. 750°F (*399°C*)
   c. 1,000°F (*538°C*)
   d. 1,250°F (*677°C*)

Identify

D. Identify the following construction types. Write the type number before each description.

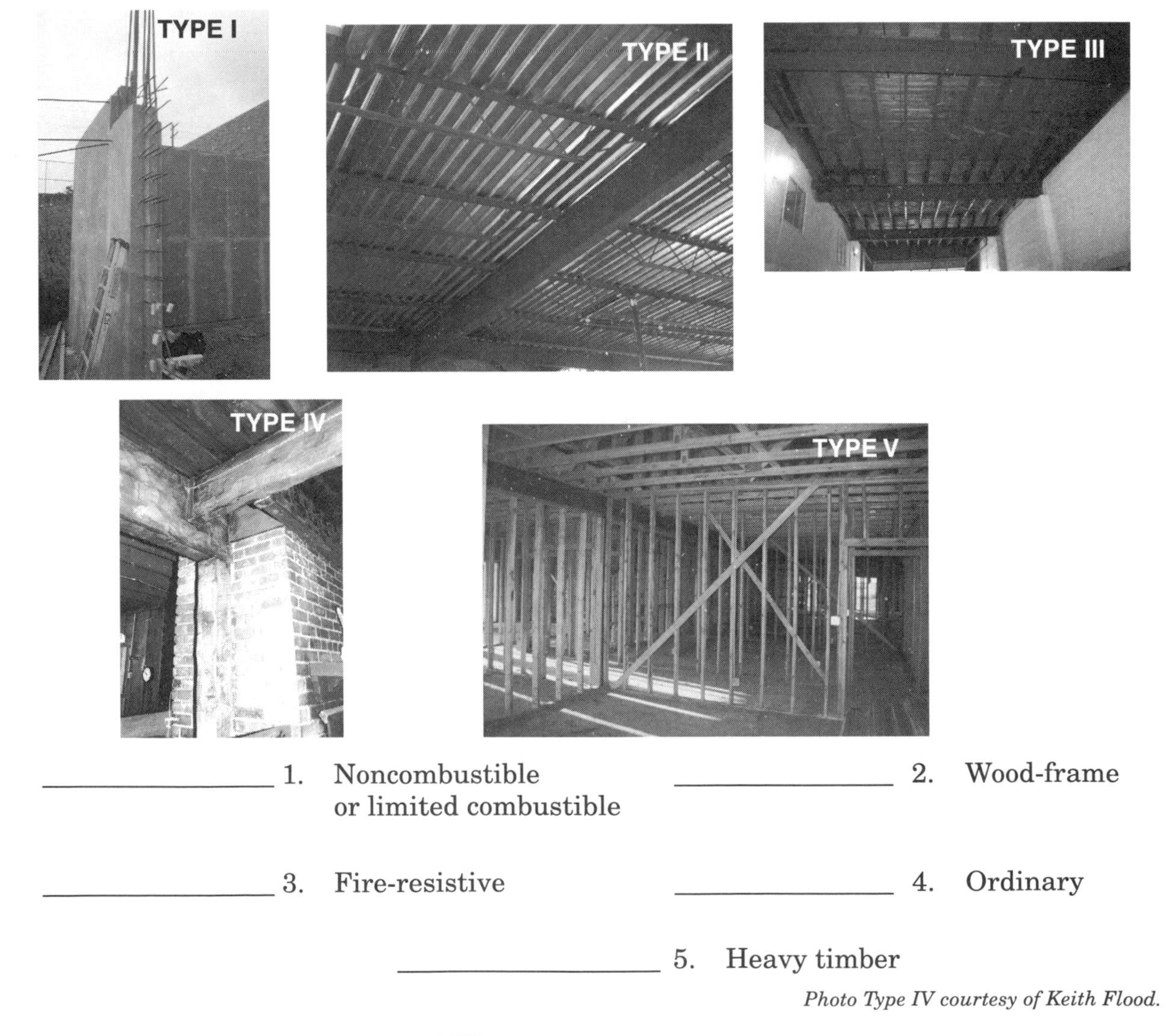

__________ 1. Noncombustible or limited combustible

__________ 2. Wood-frame

__________ 3. Fire-resistive

__________ 4. Ordinary

__________ 5. Heavy timber

Photo Type IV courtesy of Keith Flood.

FIREFIGHTER II

List

A. **List indicators of building collapse that firefighters should look for at every fire.**

1. ____
2. ____
3. ____
4. ____
5. ____
6. ____
7. ____
8. ____
9. ____
10. ____
11. ____

B. **List two fire fighting operations that increase the risk of building collapse.**

1. ____
2. ____

Chapter 4
Firefighter Personal Protective Equipment

Chapter 4 Firefighter Personal Protective Equipment

FIREFIGHTER I

Matching

A. Match to their definitions terms associated with personal protective equipment. Write the appropriate letters on the blanks.

____ 1. Wildland fire fighting personal protective clothing

____ 2. Accumulation of fluids in the lungs and associated swelling

____ 3. Suspension of small particles of carbon, tar, and dust floating in a combination of heated gases

____ 4. Hemoglobin combined with and carrying oxygen in a loose chemical combination

____ 5. Breathing apparatus that uses compressed or liquid oxygen

____ 6. Emergency breathing technique used to extend the use of the remaining air supply

a. Smoke
b. Carboxyhemoglobin
c. Brush gear
d. Oxyhemoglobin
e. Pulmonary edema
f. Skip breathing
g. Rebreather

B. Match toxic atmospheres to their characteristics. Write the appropriate letters on the blanks. Characteristics are continued on next page.

____ 1. Dark brown fuming liquid or reddish-brown gas with a pungent, acrid odor

____ 2. Colorless or pale blue liquid or gas with a bitter almond odor; may be absorbed through the skin

____ 3. Colorless, odorless gas usually indicated by black smoke

a. Carbon monoxide
b. Hydrogen chloride
c. Carbon dioxide
d. Nitrogen dioxide
e. Phosgene
f. Nitrous oxide
g. Hydrogen cyanide

____ 4. Colorless, tasteless liquid or gas with a sweet musty hay odor at low concentrations; sharp, pungent odor at higher concentrations

____ 5. Colorless to slightly yellow gas with an irritating, pungent odor

____ 6. Nonflammable, colorless, odorless gas; can be a liquid or solid

C. Match toxic atmospheres to their primary sources. Write the appropriate letters on the blanks.

____ 1. Flame contact with refrigerants such as Freon

____ 2. Present with every fire

____ 3. Heated plastics such as polyvinyl chloride (PVC) containing chlorine

____ 4. Fire extinguishing agent; common end product of complete combustion

____ 5. Fermenting grains, decomposing pyroxylin plastics

____ 6. Heated wool, nylon, polyurethane foam, rubber, paper

a. Carbon monoxide
b. Hydrogen chloride
c. Carbon dioxide
d. Nitrogen dioxide
e. Phosgene
f. Nitrous oxide
g. Hydrogen cyanide

D. Match toxic atmospheres to specific locations in which they are *most likely* to be found. Write the appropriate letters on the blanks. Locations are continued on next page.

____ 1. All fires

____ 2. Drug, toy, and general merchandise stores; telephone and electrical cables; plastics in the home

____ 3. Appliance stores, supermarkets, meat markets; other facilities containing cold storage equipment

a. Carbon monoxide
b. Hydrogen chloride
c. Carbon dioxide
d. Nitrogen dioxide
e. Phosgene
f. Nitrous oxide
g. Hydrogen cyanide

____ 4. Facilities that have CO_2 total flooding systems; facilities that manufacture wood alcohol, ethylene, dry ice, or carbonated beverages

____ 5. Clothing stores, rug shops, aircraft cabins

____ 6. Grain bins, silos, office supply stores

True/False

E. Write *True* or *False* before each of the following statements. Correct those statements that are false.

________ 1. All protective equipment has inherent limitations.

____

____

________ 2. At the emergency scene, only firefighters entering the burning building need to wear full protective equipment.

____

____

________ 3. Firefighters may remove the product label from protective clothing only after the clothing has been appropriately inventoried.

____

____

________ 4. Faceshields provide secondary protection for the face and eyes when SCBA is not required.

____

____

________ 5. Eye injuries can be serious, but they are fairly easy to prevent.

____

____

_____ 6. The hood is the only piece of personal protective gear allowed to interfere with the facepiece-to-face seal.

____

____

_____ 7. Three-quarter boots and long coats provide adequate protection for the lower torso and extremities.

____

____

_____ 8. Gloves that provide the best protection do not have the most dexterity.

____

____

_____ 9. Protective boots and safety shoes can be worn interchangeably.

____

____

_____ 10. Firefighters who wear the same size may share protective boots.

____

____

_____ 11. Firefighters should not wear synthetic materials at a fire.

____

____

_____ 12. Work uniforms are fire-resistant and designed to be worn for fire fighting operations when necessary.

____

____

_____ 13. Departments must contact the manufacturer before repainting a helmet.

____

____

_______ 14. The best way to immediately reverse tissue damage from inhaling hot air is to introduce fresh, cool air.

____

____

_______ 15. Black smoke indicates a low level of carbon monoxide.

____

____

_______ 16. It may take years for carboxyhemoglobin to dissipate from the bloodstream.

____

____

_______ 17. Firefighters do not necessarily show symptoms when suffering from carbon monoxide exposure.

____

____

_______ 18. First aid workers should administer pure oxygen to victims of carbon monoxide exposure.

____

____

_______ 19. Hazardous or toxic atmospheres are associated only with smoke produced by fire.

____

____

_______ 20. Large enclosed areas such as storage tanks, bins, or silos present no breathing hazards unless toxic gas is present.

____

____

_______ 21. Firefighters should attempt to view warning placards on overturned trucks from a distance before moving in close to the scene of the accident.

____

____

_______ 22. Open-circuit SCBA is the most commonly used protective breathing apparatus in the fire service.

____

____

_______ 23. NIOSH and MSHA approve SCBA part substitutions on interchangeable products manufactured by different companies.

____

____

_______ 24. Audible alarms on SCBA units sound when the cylinder pressure drops below one-eighth of the maximum rated pressure.

____

____

_______ 25. Firefighters should not have to remove their seat belts to don seat-mounted SCBA.

____

____

_______ 26. Firefighters should leave the SCBA facepiece connected to the regulator during storage.

____

____

_______ 27. NFPA standards approve the practice of firefighters standing to don SCBA en route to an emergency only if the firefighter is in an enclosed portion of the apparatus.

____

____

_____ 28. During inspection, the low-pressure alarm should sound briefly when the cylinder valve is turned on and again as the pressure is relieved.

_____ 29. The person who checks the bypass valve on an SCBA should leave it open to indicate that it has been inspected.

_____ 30. Firefighters usually cannot determine their own limitations and must depend on each other to watch for symptoms of heat-related conditions.

_____ 31. Improved visibility does not ensure that an area is free from contamination.

_____ 32. At times firefighters must remove SCBA facepieces to maneuver through a tight area.

Multiple Choice

F. Write the letter of the best answer on the blank before each statement.

_____ 1. Which NFPA standard covers the requirements for personal protective clothing for structural fire fighting?

a. 1971
b. 1972
c. 1973
d. 1974

_____ 2. To comply with NFPA standards, what information should be included on the labels of protective clothing?

a. Expiration date
b. Month and year of manufacture
c. Garment inspector's name
d. Garment inspection date

____ 3. Why do firefighter helmets have wide brims?
- a. To vent smoke away from the firefighter's eyes and nose
- b. To provide protection from hot air drafts
- c. To prevent embers from reaching the ears and neck
- d. To supply secondary protection for the face and ears

____ 4. Which component must all structural fire fighting helmets have?
- a. 3-inch *(75 mm)* visor
- b. Flashlight attachment
- c. Velcro® fasteners
- d. Ear covers

____ 5. What eye protection may *not* provide the safeguard required for flying particles or splashes?
- a. Goggles
- b. Faceshields
- c. Safety glasses
- d. SCBA masks

____ 6. How should departments manage the use of eye protection?
- a. Present the option in departmental memos.
- b. Demonstrate the need during special training sessions.
- c. Stress the importance through verbal recommendations.
- d. Require use through standard operating procedures.

____ 7. In which of the following situations are safety glasses or goggles most appropriate?
- a. Vehicle extrication
- b. Structural fire attack
- c. Pre-incident exterior survey
- d. Primary size-up

____ 8. Firefighters who wear prescription safety eyeglasses must select frames and lenses that meet ____.
- a. OSHA 221.1
- b. NFPA 71
- c. ANSI Z87.1
- d. IFSTA 44.8

____ 9. How should earplugs be distributed?
- a. Every firefighter should have a personal set.
- b. Units should keep several sets on hand for firefighters to check out.
- c. Fire departments should keep used sets on hand for anyone to use.
- d. Fire departments should ban the use of earplugs.

____ 10. How does ear protection affect structural fire fighting procedures?
- a. Earmuffs fit well under most SCBA gear.
- b. Earplugs are safe in any structural fire attack.
- c. Earmuffs shut out too much noise and make firefighters feel disoriented.
- d. Earplugs may melt when exposed to intense heat.

____ 11. According to NFPA 1971, how many layers must protective coats have?

a. Two
b. Three
c. Four
d. Five

____ 12. In what emergency situation can a firefighter remove the inner liners of a protective coat?

a. During structural fire fighting in the heat of the summer
b. During search and rescue at a fire scene
c. After a fire has died down to the decay stage
d. Never

____ 13. What prevents water or fire products from entering a protective coat through the gaps between snaps or clips?

a. Nothing
b. Reflector strips
c. Closure system
d. Liners

____ 14. How should protective trousers be constructed?

a. With the same number of layers as protective coats plus an abrasion-resistant inner shell
b. With the same number of layers as protective coats
c. With the same number of layers as protective coats but without a moisture barrier
d. With the same number of layers as protective coats but without a thermal barrier

____ 15. Where should the puncture-resistant stainless steel plate be located in protective boots?

a. Midsole
b. Toe
c. Full-sole
d. Heel

____ 16. Which NFPA standard addresses personal protective clothing for wildland fire fighting?

a. 1971
b. 1973
c. 1975
d. 1977

____ 17. What underwear should be worn under brush gear?

a. 60 percent silk, including short-sleeved T-shirt, and nylon-blend socks
b. 100 percent silk, including a long-sleeved T-shirt, and nylon-blend socks
c. 60 percent cotton, including long-sleeved T-shirt, and natural-fiber socks
d. 100 percent cotton, including long-sleeved T-shirt, and natural-fiber socks

____ 18. Are synthetic materials recommended for fire fighting operations? Why or why not?

a. Yes. They do not readily absorb water, which reduces the occurrence of firefighter fatigue.
b. Yes. They insulate the firefighter from heat exposure.
c. No. They provide too much insulation, which prevents the firefighter from accurately predicting critical temperatures.
d. No. They melt when heated and stick to the wearer's skin.

____ 19. What head and face protection is preferred for wildland fire fighting?

a. Lightweight helmet and goggles
b. Structural helmet and goggles
c. Structural helmet with faceshield
d. Lightweight helmet with faceshield

____ 20. What standard guidelines apply to boots for wildland fire fighting in all areas?

a. Wellington-style boots with lug- or grip-tread soles, 8 to 10 inches *(200 mm to 250 mm)* high
b. Lace-up or zip-up safety boots with smooth rubber soles, 10 to 12 inches *(250 mm to 300 mm)* high
c. Lace-up or zip-up safety boots with lug- or grip-tread soles, 8 to 10 inches *(200 mm to 250 mm)* high
d. Wellington-style boots with lug- or grip-tread soles, 10 to 12 inches *(250 mm to 300 mm)* high

____ 21. Which of the following fabrics has a low temperature resistance and should be avoided for station/work uniforms?

a. Nylon
b. Wool
c. Neoprene
d. Cotton

____ 22. According to NFPA standards, at what heat and for how long should components of station/work uniforms withstand ignition, melting, dripping, or separating?

a. 300°F *(149°C)* for 3 minutes
b. 400°F *(204°C)* for 4 minutes
c. 500°F *(260°C)* for 5 minutes
d. 600°F *(316°C)* for 6 minutes

____ 23. How does dirt affect the protection provided by a helmet?

a. Absorbs heat faster than the shell itself
b. Softens the shell material and reduces its impact and dielectric protection
c. Reduces the helmet's ability to resist the transmission of force
d. Deteriorates the polycarbonate material and makes the helmet brittle

____ 24. According to NFPA 1500, how should fire departments clean protective clothing?

a. They should be taken to a Laundromat with heavy-load-capacity washers and dryers.

b. They should be sent to a cleaning service.

c. They should be washed in the same facilities that firefighters use for their daily-wear clothes.

d. They should be hand washed at the station.

____ 25. According to NFPA 1581, how often should personal protective clothing be cleaned and dried?

a. Once every month

b. At least every two months

c. At least every six months

d. Once per year

____ 26. Which body areas are *most* vulnerable to injury during a fire attack?

a. Eyes and internal ear canals

b. Fingers and wrists

c. Lungs and respiratory tract

d. Lower back and hamstring muscles

____ 27. When oxygen in the atmosphere drops below ____ percent, the human body responds by increasing the respiratory rate.

a. 15

b. 18

c. 21

d. 24

____ 28. How does combining irritants and toxicants change the effects that each has on the human body?

a. They become more toxic.

b. They react with each other and become unable to interact with human tissue.

c. They dissipate more quickly.

d. They release fewer toxicants.

____ 29. Which toxic gas forms hydrochloric acid in the lungs?

a. Nitrogen dioxide

b. Ammonia

c. Hydrogen cyanide

d. Phosgene

____ 30. How does hydrogen chloride gas attack the body?

a. Combines with the blood's hemoglobin and crowds oxygen from the blood

b. Causes swelling and obstruction of the upper respiratory tract

c. Hampers respiration at the cellular and tissue level

d. Paralyzes the brain's respiratory center

____ 31. What determines how deeply a smoke particle will be inhaled into an unprotected lung?

a. Temperature of the particle
b. Size of the particle
c. Capacity of the lung
d. Density of the smoke

____ 32. With a room air concentration of 1 percent carbon monoxide, how much will the carbon monoxide level in the blood elevate within 2 to 7 minutes?

a. 25 percent
b. 35 percent
c. 50 percent
d. 75 percent

____ 33. Which of the gases below has an almond odor?

a. Nitrogen oxide
b. Hydrogen chloride
c. Hydrogen cyanide
d. Phosgene

____ 34. What is the IDLH of carbon dioxide?

a. 40,000 ppm
b. 50,000 ppm
c. 60,000 ppm
d. 70,000 ppm

____ 35. What is the IDLH of carbon monoxide?

a. 500 ppm
b. 1,200 ppm
c. 5,000 ppm
d. 12,500 ppm

____ 36. How does carbon monoxide attack the body?

a. Combines with the blood's hemoglobin and crowds oxygen from the blood
b. Causes swelling and obstruction of the upper respiratory tract
c. Interferes with respiration at the cellular and tissue level by hampering the proper exchange of oxygen and carbon dioxide
d. Paralyzes the brain's respiratory center

____ 37. Which of the following gases has a musty hay odor and forms hydrochloric acid in the lungs when inhaled?

a. Nitrogen oxide
b. Nitrogen chloride
c. Phosgene
d. Hydrogen cyanide

____ 38. What is the DOT definition of a hazardous material?

a. Any substance that poses probable risk to health and safety or property when exposed to heat or solar energy
b. Any substance that is likely to inflict injury or harm or impose great or continued risk unless dealt with carefully
c. Any substance which may pose an unreasonable risk to health and safety if not properly controlled during handling
d. Any substance that may pose a hazard to humans, wildlife, or the environment on contact

____ 39. To which substance can firefighters develop a tolerance so that they suffer its harmful effects without symptoms?

a. Carbon monoxide
b. Hydrogen cyanide
c. Phosgene
d. Sulfur dioxide

____ 40. Which of the following scenarios is true about the presence of carbon monoxide during a fire?

a. The higher the fire temperature, the less carbon monoxide is present
b. The better the ventilation, the greater the quantity of carbon monoxide
c. The more efficient the burning, the greater the quantity of carbon monoxide
d. The darker the smoke, the higher the level of carbon monoxide

____ 41. At approximately what rate does carbon monoxide combine with hemoglobin in the bloodstream?

a. 100 times more readily than does oxygen
b. 200 times more readily than does oxygen
c. 100 times less readily than does oxygen
d. 200 times less readily than does oxygen

____ 42. At what concentration of carbon monoxide does an atmosphere *become* dangerous?

a. 0.0005 percent (5 ppm)
b. 0.005 percent (50 ppm)
c. 0.05 percent (500 ppm)
d. 0.5 percent (5000 ppm)

____ 43. At what concentration in the air can carbon monoxide cause unconsciousness and death?

a. 0.01 percent
b. 0.02 percent
c. 0.25 percent
d. 1.01 percent

____ 44. Why is it dangerous for a firefighter to reenter a smoky atmosphere after quickly recovering from the effects of carbon monoxide exposure?

a. The signs of nerve or brain injury may appear any time within three weeks.
b. The experience may make the firefighter too apprehensive to continue successfully.
c. The chances of being overcome by carbon monoxide while wearing SCBA increase with every exposure.
d. The firefighter may develop a dangerous level of resistance to carbon monoxide.

____ 45. Which standard allows firefighters to wear soft contact lenses while using SCBA after demonstrating successful long-term (at least 6 months) use of contact lenses without any problems?

a. NFPA 1500
b. ANSI Z87.1
c. 29 CFR 1910.134
d. OSHA 221.1

____ 46. To what part of the body do waist straps distribute the weight of SCBA cylinders or packs?

a. Shoulders
b. Hips
c. Knees
d. Upper back

____ 47. What action voids the NIOSH and MSHA warranties on SCBA units?

a. Removing waist straps
b. Refilling air cylinders
c. Bypassing failed regulator valve controls
d. Using a side- or rear-mount donning method

____ 48. How many cubic feet *(liters)* of breathing air does a fully filled typical 30-minute-rated, 2,216 psi *(15 290 kPa)* air cylinder contain?

a. 30 *(850)*
b. 35 *(990)*
c. 40 *(1 130)*
d. 45 *(1 270)*

____ 49. In what situation should a firefighter use an SCBA bypass valve?

a. The firefighter begins to hyperventilate.
b. The firefighter needs to control airflow to continue working after the low-pressure alarm sounds.
c. The regulator valve fails.
d. The firefighter panics and wants to manually control the airflow.

____ 50. The remote pressure gauge should read within ____ psi (*kPa)* of the cylinder gauge if the increments are in psi.

a. 100 *(700)*
b. 200 *(1 400)*
c. 300 *(2 100)*
d. 400 *(2 800)*

____ 51. What should teams do if one team member's low-pressure alarm sounds during the suppression of a fire?

a. The team member with air should share his/her air supply with the team member who is low on air.
b. The team member low on air should switch to the bypass valve to continue working for up to 15 minutes.
c. The team member who is low on air should leave the area immediately after informing the partner.
d. Both members of the team should stop working and exit the fire area immediately.

____ 52. Which of the following methods is recommended for preventing or controlling internal fogging on the SCBA mask?

a. Releasing cylinder air
b. Exhaling forcefully
c. Using a nosecup
d. Applying toothpaste

____ 53. What is the benefit of using closed-circuit rather than open-circuit SCBA during hazardous materials incidents?

a. Provides longer air-supply duration
b. Uses compressed air
c. Vents exhaled air out of the system
d. Does not require as much training

____ 54. Approximately how many seconds must a firefighter remain motionless before a PASS device emits a loud, pulsating shriek?

a. 15
b. 30
c. 45
d. 90

____ 55. When should a firefighter turn on and test a PASS device?

a. Prior to donning the equipment
b. Just before entering a burning structure
c. After determining a possible need for the device
d. Before entering each new fire-involved room

____ 56. How often should firefighters retrain in the use of PASS devices?

a. Monthly
b. Quarterly
c. Every six months
d. Yearly

____ 57. At what percentage of cylinder capacity does NFPA 1404 consider a cylinder to be full?

a. No less than 85 percent
b. No less than 90 percent
c. No less than 95 percent
d. No less than 97 percent

____ 58. Which of the following mounts permits donning SCBA en route?

a. Side mount
b. Compartment mount
c. Backup mount
d. Seat mount

____ 59. What is the best way for a firefighter to ensure a proper facepiece-to-face seal?

a. Cutting long hair above the ears or pulling it back into a ponytail and using a standard-sized facepiece
b. Tightening facepiece straps as tight as they will comfortably tighten without cutting off circulation to the face
c. Being fitted with a facepiece and preventing hair or other obstacles from getting between the skin and the facepiece
d. Removing nosecups and eyeglasses and tightening the facepiece straps securely

____ 60. What is the best way to test a facepiece for positive pressure?
   a. Gently break the facepiece seal by inserting two fingers under the edge of the facepiece; feel for air to gently move past your fingers.
   b. While donning the facepiece, place two fingers at each side of the facepiece; feel for air to move gently past your fingers.
   c. Before donning the facepiece, place two fingers at each side of the facepiece and hold it to the face; feel for air to move gently past your fingers.
   d. While testing the facepiece, hold one hand over the facepiece and pressurize the unit; feel for air to move gently past your fingers.

____ 61. According to NFPA 1404 and NFPA 1500, how often should SCBA be inspected?
   a. After each use, annually, and randomly
   b. Monthly, annually, and randomly
   c. Weekly, monthly, annually, and randomly
   d. After each use, weekly, monthly, and annually

____ 62. How should a facepiece be washed?
   a. With warm water and mild disinfectant
   b. With cold water and dishwashing detergent
   c. With hot water and a strong detergent
   d. With an autoclave

____ 63. With what should a facepiece be dried?
   a. Lint-free cloth
   b. Terry cloth towel
   c. Paper towel
   d. Tissue paper

____ 64. Who should perform annual inspections of SCBA?
   a. Firefighters
   b. Equipment supervisors
   c. Safety officers
   d. Certified technicians

____ 65. How often does DOT require that steel and aluminum cylinders be tested?
   a. Every 5 years
   b. Every 4 years
   c. Every 3 years
   d. Every 2 years

____ 66. How often does DOT require that composite cylinders be tested?
   a. Every 5 years
   b. Every 4 years
   c. Every 3 years
   d. Every 2 years

____ 67. Which precaution applies to filling any air cylinder?
   a. Put the cylinder into an unshielded charging station.
   b. Avoid chatter in the cylinder by filling slowly.
   c. Overpressurize the cylinder and then release excess pressure to ensure a complete fill.
   d. Prevent overheating by filling the cylinder as quickly as possible.

____ 68. Which of the following methods is *not* an approved procedure for finding one's way from a burning structure in an emergency?
   a. Find the nearest stairway and follow the handrail down.
   b. Break a window or breach a wall.
   c. Contact a wall and crawl in one direction only.
   d. Activate the PASS device and call out for help.

____ 69. How does the skip breathing technique work?
   a. The firefighter exhales twice as long as normal and then inhales a regular breath.
   b. The firefighter holds a normal breath for ten seconds and then exhales all of the air from the lungs.
   c. The firefighter inhales as much air as possible into the lungs, holds the breath for three seconds, and then exhales all of the air during a five-second count.
   d. The firefighter inhales a normal breath, holds the breath as long as it would take to exhale, and then inhales once again before exhaling.

____ 70. What is the primary method of moving about in areas of obscured visibility?
   a. Walking with arms extended
   b. Walking bent over at the waist
   c. Crouching or "duck" walking
   d. Crawling

____ 71. What should the firefighter do with SCBA cylinders that are out of hydrostatic test date?
   a. Discard them to be recycled.
   b. Overhaul them in the fire department shop, and then send them to the manufacturer for hydrostatic testing.
   c. Remove them from service, and tag them for further inspection and hydrostatic testing.
   d. Inspect and hydrostatically test them, and then place them back in service.

Identify

G. Identify the following abbreviations associated with firefighter personal protective equipment. Write the correct interpretation before each.

________________________________ 1. SCBA

________________________________ 2. PASS

_______________ 3. IDLH

_______________ 4. NIOSH

_______________ 5. CO

_______________ 6. COHb

_______________ 7. ppm

_______________ 8. DOT

_______________ 9. OSHA

_______________ 10. MSHA

_______________ 11. PAD

H. Identify the correct order for doffing SCBA. Number the following steps in order from first (1) to last (7). Write the correct numbers on the blanks next to each.

____ a. Close cylinder valve.

____ b. Disconnect the low-pressure hose from the regulator or remove the regulator from the facepiece, depending upon type of SCBA.

____ c. Discontinue the flow of air from the regulator to the facepiece.

____ d. Remove the backpack assembly while protecting the regulator.

____ e. Extend all straps.

____ f. Relieve pressure from the regulator in accordance with manufacturer's instruction.

____ g. Remove the facepiece.

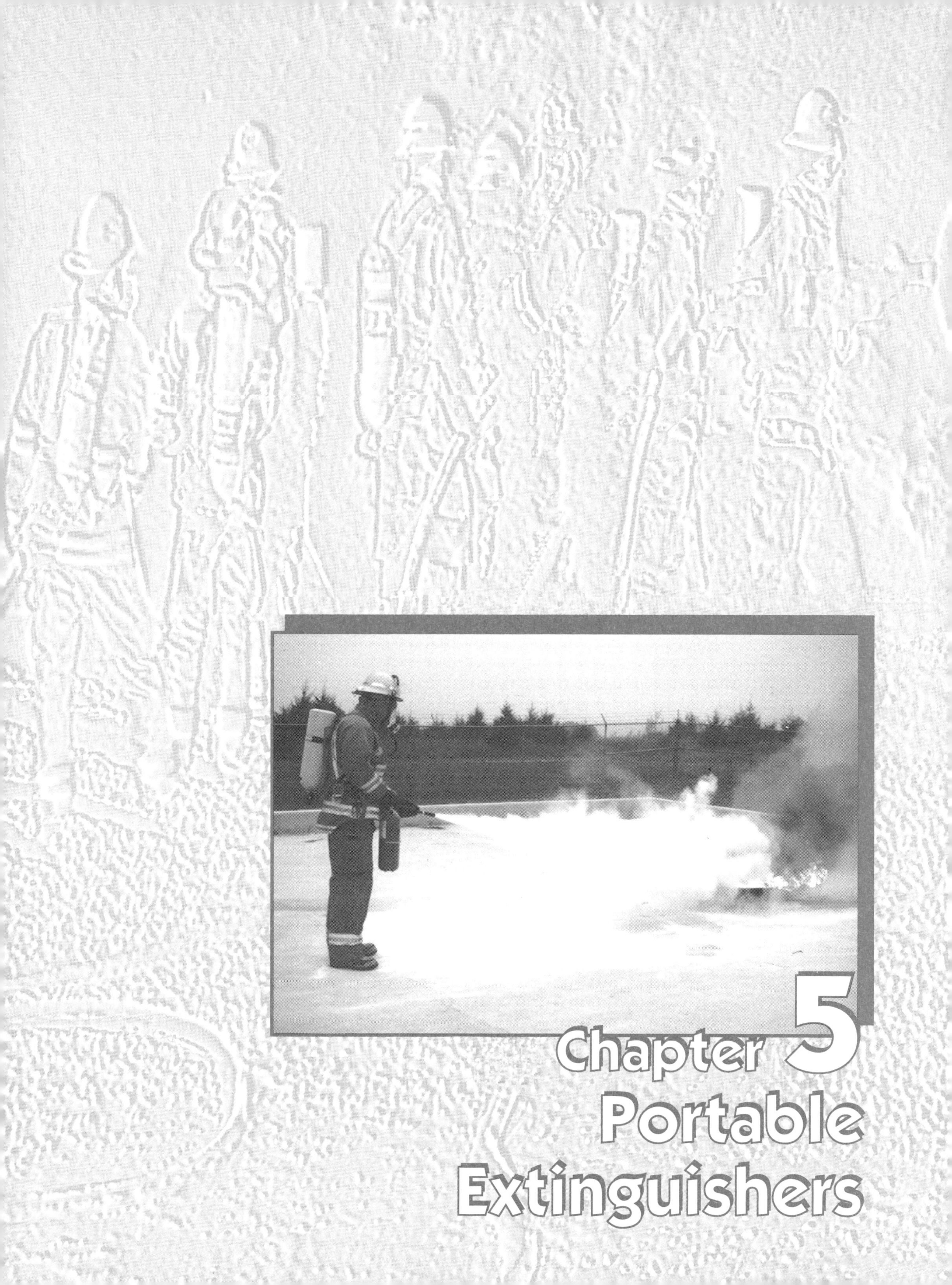

Chapter 5
Portable Extinguishers

Chapter 5 Portable Extinguishers

FIREFIGHTER I

Matching

A. Match the types of portable fire extinguishers to their primary uses. Write the appropriate letters on the blanks.

____ 1. For small Class A fires only

____ 2. For all types of small Class A fires, confined hot spots during overhaul operations, and chimney flue fires

____ 3. For Class A and Class B fires, particularly small liquid fuel spills

____ 4. For most total flooding systems

____ 5. For Class D fires only

____ 6. For Class B and Class C fires, but with limited reach

____ 7. For use on Class A-B-C fires and/or Class B-C fires

a. AFFF
b. Halon 1211
c. Pump-tank water
d. Halon 1301
e. CO_2
f. Dry chemical
g. APW
h. Dry powder

True/False

B. Write *True* or *False* before each of the following statements. Correct those statements that are false.

________ 1. Never rely on extinguishers found in occupancies.

____

____

________ 2. According to NFPA standards, pumping apparatus must carry at least one approved portable fire extinguisher with mounting bracket.

____

____

_______ 3. The AFFF extinguisher has an air aspirating nozzle that aerates the foam solution, producing a better quality foam than a standard nozzl

_______ 4. Never add antifreeze to water-type extinguishers; instead, store extinguishers in a warm place to prevent freezing.

_______ 5. Mixing AFFF and water creates a vapor seal that is effective on fires involving alcohol and acetone.

_______ 6. By the year 2000, all halogen manufacturers must phase out halogen products except those products that do not have suitable alternatives

_______ 7. Halogenated vapor does not effectively extinguish surface fires in flammable and combustible liquids.

_______ 8. Halon decomposes and liberates toxic components, so it should not be used in unventilated, confined spaces.

_______ 9. Carbon dioxide is stored under its own pressure as a liquefied compressed gas ready for use at any time.

_______ 10. Carbon dioxide produces a vapor-suppressing film on the surface of fuel, eliminating the danger of the fuel reigniting.

____

____

_______ 11. Class D fires should be extinguished with either dry chemical agents or dry powder agents.

____

____

_______ 12. Never mix dry chemicals with any other type of agent because they may chemically react and cause a dangerous rise in pressure inside the extinguisher.

____

____

_______ 13. Dry chemicals, generally considered toxic, reduce visibility and create respiratory problems.

____

____

_______ 14. No single agent will control or extinguish fires in all combustible metals.

____

____

_______ 15. When fighting a fire on combustible metal, apply a sufficient amount of dry powder to create a smothering blanket and be careful not to break any crust that forms during this process.

____

____

_______ 16. Manufacturers use number ratings to identify extinguishers suitable for more than one class of fire.

____

____

_______ 17. The ratings for each fire class are independent and do not affect each other.

____

____

_______ 18. Always select extinguishers that maximize fire control regardless of the risk to property.

____

____

_______ 19. Firefighters need a good working knowledge of how to operate fire extinguishers, but they do not need to become familiar with the detailed instructions of every extinguisher.

____

____

_______ 20. Place empty fire extinguishers back on the apparatus to reduce the chance of someone else trying to use the empty extinguishers on the fire.

____

____

_______ 21. Fire inspectors should include in their pre-incident planning programs the examination of portable extinguishers during building inspections.

____

____

_______ 22. Never attempt to repair the shell or cylinder of a defective extinguisher.

____

____

Multiple Choice

C. Write the letter of the best answer on the blank before each statement.

___ 1. What information does NFPA 10 provide in regard to portable fire extinguishers?

a. Use, placement, and size
b. Placement, size, and rating
c. Size, rating, and use
d. Rating, placement, and use

____ 2. What is the minimum rating for a carbon dioxide extinguisher carried on a pumping apparatus?

a. 10 B:C
b. 30 B:C
c. 50 B:C
d. 70 B:C

____ 3. How may water-type extinguishers be protected from freezing?

a. Add antifreeze to the water.
b. Use only water from the pumper.
c. Keep them on the apparatus at all times.
d. Remove the water from the extinguisher during cold spells.

____ 4. Which of the following statements best describes pump-tank water extinguishers?

a. When the operating valve is activated, the water is forced up the siphon tube and out through the hose.
b. They are generally equipped with a double-acting pump.
c. The agent is expelled by compressed nitrogen stored in the tank.
d. A gauge located on the side of the valve assembly shows when the extinguisher is properly pressurized.

____ 5. Which foam serves as a wetting agent that aids in extinguishing deep-seated fires and vehicle fires?

a. Class A
b. Class B
c. Class C
d. Class D

____ 6. What precaution should be taken when applying foam?

a. Prevent foam from gently raining down onto fuel.
b. Avoid deflecting foam off an object.
c. Do not apply foam directly onto fuel.
d. Apply foam directly onto fuel.

____ 7. In which situation should an AFFF extinguisher be used?

a. On Class C fuels
b. On Class D fuels
c. On three-dimensional fires
d. On static pools of flammable liquids

____ 8. Halons are most effective on fires in which of the following types of fuel?

a. Self-oxidizing fuels
b. Sensitive electronic equipment
c. Organic peroxides
d. Metal hydrides

____ 9. Which extinguisher requires freeze protection?

a. Halon 1211
b. CO_2
c. APW
d. Halon 1301

____ 10. What are the two basic types of dry chemical extinguishers?

a. Regular B:C-rated; multipurpose and A:B:C-rated
b. Multipurpose A-rated; multipurpose B:C-rated
c. Regular B:C-rated; regular A:B:C-rated
d. Multipurpose B:C-rated; regular A:B:C-rated

____ 11. Why do manufacturers mix additives with dry chemicals?

a. To dilute the toxicity of the chemicals
b. To reduce the particulate per oxygen ratio of the chemicals
c. To suppress the corrosive tendencies of the chemicals
d. To prevent moisture from caking the chemicals

____ 12. What rule should be followed when pressurizing a wheeled dry chemical extinguisher?

a. Keep your head directly above the top of the unit.
b. Be prepared for a significant nozzle reaction.
c. Remove the hose after the extinguisher has been charged.
d. Have another firefighter hold the nozzle firmly.

____ 13. How should firefighters extinguish burning metal on a combustible surface and reduce reignition risks?

a. Cover the fire with powder; then move the burning metal onto a 1- to 2-inch *(25 mm to 50 mm)* layer of powder.
b. Move the burning metal onto a 1- to 2-inch *(25 mm to 50 mm)* layer of powder; then cover the combustible surface with powder.
c. Cover the fire with a 2- to 4-inch *(50 mm to 100 mm)* layer of powder; then monitor the fire for five hours.
d. Move the burning metal onto a 2- to 4-inch *(50 mm to 100 mm)* layer of powder; then leave the metal and the combustible surface undisturbed until both have completely cooled.

____ 14. What determines a Class A rating for a fire extinguisher?

a. The approximate square foot *(square meter)* area of a fire that a nonexpert can extinguish
b. The amount of extinguishing agent and the duration and range of the discharge
c. The relative conductivity of the agent
d. The toxicity of the fumes produced and the products of combustion

____ 15. What information can be inferred from a rating of 4-A 20-B:C on a multipurpose extinguisher?

a. It is not safe to use on fires involving energized electrical equipment.

b. It can extinguish approximately one-fourth as much Class A fire as a Class 1-A extinguisher.

c. It can extinguish a deep layer of flammable liquid fire of a 20 square-foot *(2 m^2)* area.

d. It can extinguish a Class B fire two times larger than a 1-B extinguisher can.

____ 16. What should a firefighter check immediately before using any extinguisher?

a. The pressure gauge

b. The instructions for use

c. The expiration date

d. All of the above

____ 17. What should a firefighter do if a fire is not extinguished after an entire extinguisher has been discharged?

a. Withdraw and reassess the situation.

b. Immediately grab another extinguisher and continue attacking the fire.

c. Withdraw and overhaul the fireground with the appropriate tool.

d. Immediately charge a hoseline and resume the attack.

____ 18. What three factors determine the value of a fire extinguisher?

a. Size, rating, and accessibility

b. Rating, serviceability, and hydrostatic test results

c. User's ability to operate it, size, and rating

d. Serviceability, accessibility, and user's ability to operate it

____ 19. Which standard regulates the procedures for hydrostatic testing of fire extinguishers?

a. NFPA 10

b. NFPA 1901

c. 29 CRF 1910.157

d. NFPA 1903

____ 20. What repairs can firefighters make to damaged extinguishers?

a. Patch leaking shells or cylinders.

b. Clean corrosion from cylinders.

c. Replace leaking hoses and nozzles.

d. Repair dented shells.

Identify

D. Identify the following abbreviations associated with portable extinguishers. Write th correct interpretation before each.

_______________ 1. CO_2

_______________ 2. APW

_______________ 3. AFFF

_______________ 4. Halon

_______________ 5. UL

_______________ 6. ULC

E. Identify the procedures for inspecting portable extinguishers. Write an *X* next to eac correct procedure.

____ 1. Check to ensure that the extinguisher is accessible.

____ 2. Inspect the nozzle or horn for obstructions, cracks, dirt, or grease.

____ 3. Check the extinguisher shell for paint chipping.

____ 4. Check the operating instructions and nameplate for legibility.

____ 5. Determine that the extinguisher is at least half-full of agent.

____ 6. Check the inspection tag for the date of the previous inspection, maintenance, or recharging.

F. **Identify the alphabetical/geometric symbols for each classification of fire. Write the correct fire class below each.**

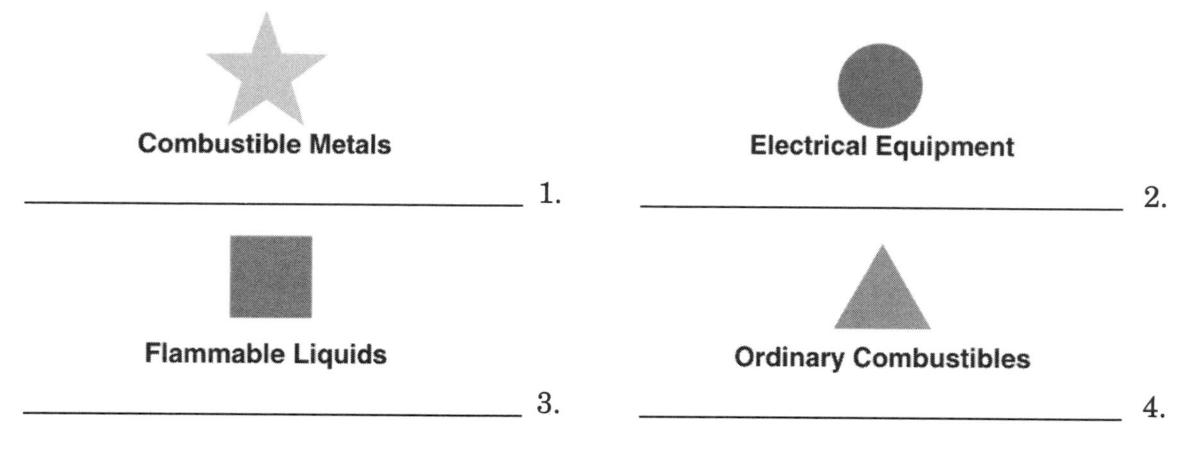

G. **Identify the pictographs representing fire classifications. Write the correct fire class below each.**

H. **Identify the fire extinguishers pictured below. Write the correct name below each.**

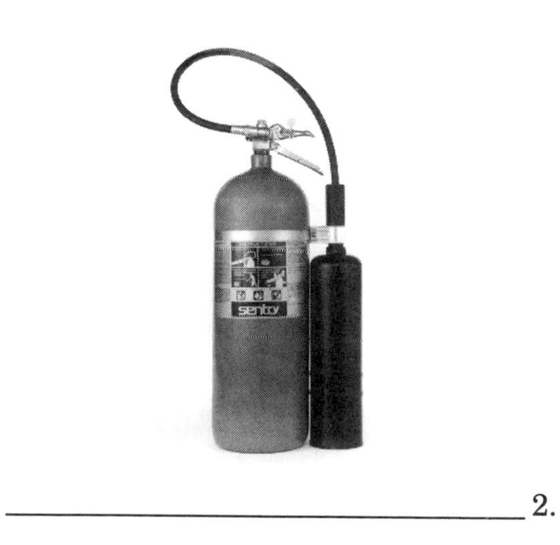

____________________ 1. ____________________ 2.

______________________  3.

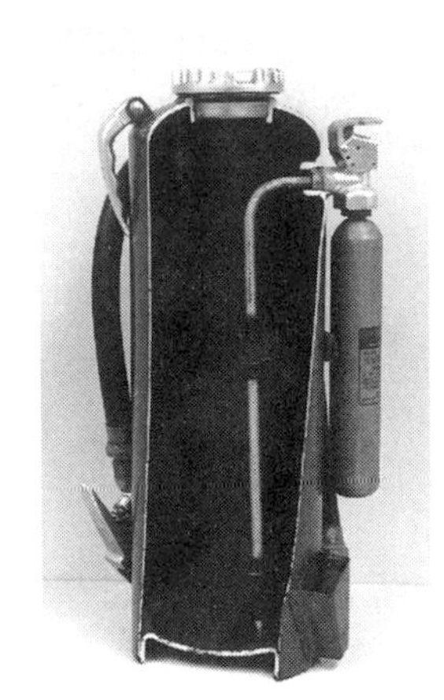

______________________  4.

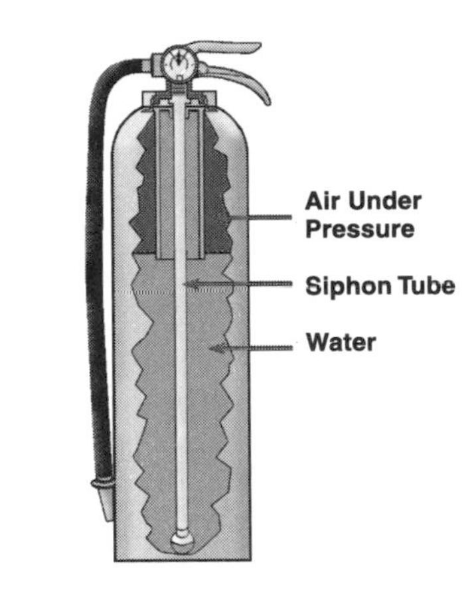

______________________ 5.

______________________  6.

______________________ 7.

______________________ 8.

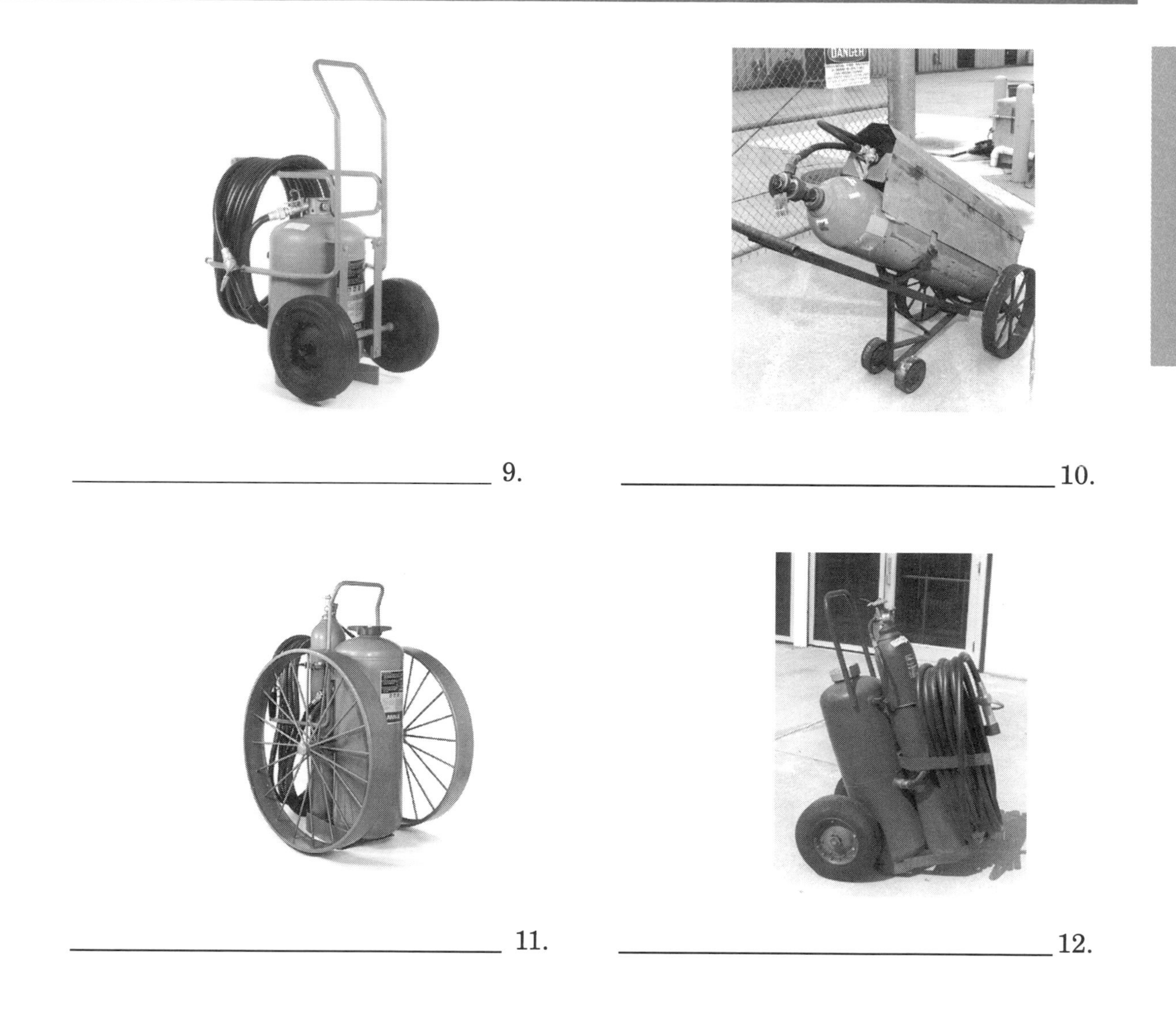

_______________ 9.

_______________ 10.

_______________ 11.

_______________ 12.

Photos for Questions 2, 9, and 11 courtesy of Ansul, Inc., Marinette, Wisconsin.
Photo for Question 10 courtesy of Conoco, Inc.

Chapter 6
Ropes and Knots

Chapter 6 Ropes and Knots

FIREFIGHTER I

Matching

A. Match to their definitions terms associated with ropes and knots. Write the appropriate letters on the blanks.

____ 1. Rope designed for high stretch without breaking

____ 2. Rope type used for hauling, rescue, and rappelling where no falls are likely to occur

____ 3. Rope end used for hoisting, pulling, or belaying

____ 4. Rope end used in forming a knot

____ 5. Rope segment between the working end and the running end

____ 6. Rope formation made by crossing the side of a bight over the standing part

a. Static
b. Loop
c. Standing part
d. Dynamic
e. Round turn
f. Bitter end
g. Running end

B. Match knots to their primary applications. Write the appropriate letters on the blanks.

____ 1. Joining ropes of unequal diameters and joining rope to chain

____ 2. Joining ropes of equal diameters

____ 3. Securing a loop in a rope for a safety line, safety harness, litter and rescue equipment, or anchor lines

____ 4. Foundation knot for other knots in family

____ 5. Safety backup

____ 6. Attaching ropes to objects and hoisting (with overhand knot)

____ 7. Forming a loop that will not constrict the object it is placed around

a. Overhand
b. Figure-eight
c. Bowline
d. Clove hitch
e. Becket bend or sheet bend
f. Figure-eight on a bight
g. Figure-eight follow through
h. Sheepshank

True/False

C. Write *True* or *False* before each of the following statements. Correct those statements that are false.

_______ 1. Utility rope can be used to support rescuers and/or victims during training evolutions.

____

____

_______ 2. Firefighters should use only rope of block creel construction (without knots or splices in the fibers) for life safety applications.

____

____

_______ 3. If rope that has been impact loaded passes inspection, it does not need to be destroyed.

____

____

_______ 4. Synthetic rope succumbs easily to mildew and rotting.

____

____

_______ 5. The construction of a laid rope leaves the load-bearing strands exposed; consequently, any damage immediately affects the rope's strength.

____

____

_______ 6. The strength of a braid-on-braid rope comes primarily from the core.

____

____

_______ 7. Ropes should be inspected both visually and tactilely after each use.

____

____

_____ 8. Mildew on synthetic laid rope indicates that the rope is no longer suitable for use.

_____ 9. After using a piece of rescue rope the first time, firefighters should start a logbook for that rope.

_____ 10. Some synthetic rope may feel stiff after washing, but this is not a cause for concern.

_____ 11. Specialized commercial rope-washing devices that connect to a standard faucet or garden hose clean rope more thoroughly than a clothes-washing machine.

_____ 12. At a fire or rescue emergency, an improperly coiled rope may result in the failure of an evolution.

_____ 13. To prevent a knot from slipping, safety knots should be tied at the working end of the rope.

_____ 14. Rope strength increases as the bends created by knots become tighter.

_____ 15. The becket bend is likely to slip on wet rope.

_______ 16. A tag line helps prevent hoisted equipment from coming into contact with other objects as it is raised.

____

____

_______ 17. Hoisting hose should be the last resort as a means for getting hoselines from the ground to upper levels.

____

____

_______ 18. Rope rescue is a basic skill that can be performed by any firefighter with Level I training.

____

____

Multiple Choice

D. Write the letter of the best answer on the blank before each statement.

____ 1. Into what two primary use classifications does fire service rope fall?

a. Standard and rescue
b. Braided and woven
c. Life safety and utility
d. Safe and unsafe

____ 2. How can firefighters determine that rope has been impact loaded?

a. Inspecting the rope thoroughly
b. Reading an entry in the rope's logbook
c. Washing the rope meticulously and looking for irregularities
d. Storing the rope in a special bag set aside for impact-loaded rope

____ 3. Which of the following tasks should *not* be performed with utility rope?

a. Hoisting equipment
b. Securing unstable objects
c. Cordoning off an area
d. Rappelling from an elevation

____ 4. What two basic categories of materials are used to construct fire service rope?

a. Fire-resistant and fireproof fibers
b. Woven and braided fibers
c. Continuous filament and chain filament fibers
d. Natural and synthetic fibers

____ 5. Which of the following rope materials is safest for rescue operations?

a. Nylon
b. Cotton
c. Manila
d. Sisal

____ 6. What type of rope is best suited to hauling heavy loads?

a. Dynamic
b. Static
c. Resistant
d. Resistive

____ 7. Which rope construction is most susceptible to abrasion and other physical damage?

a. Braided
b. Woven
c. Solid
d. Twisted

____ 8. What type of rope construction is illustrated below?

a. Laid
b. Kernmantle
c. Braided
d. Braid-on-braid

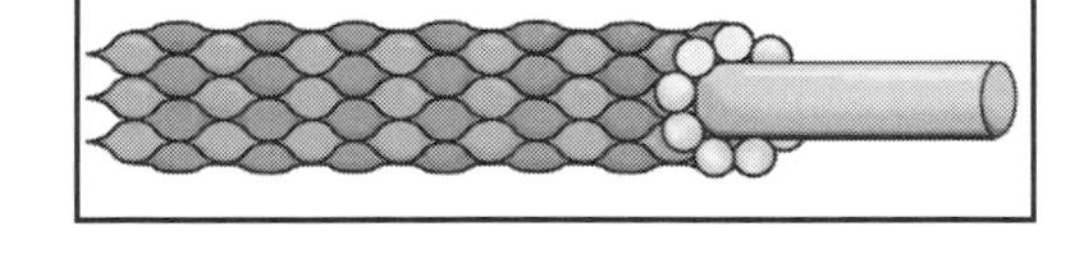

____ 9. What type of rope construction is illustrated below?

a. Laid
b. Kernmantle
c. Braided
d. Braid-on-braid

____ 10. What type of rope construction is illustrated below?

a. Laid
b. Kernmantle
c. Braided
d. Braid-on-braid

____ 11. What type of rope construction is illustrated below?

a. Laid
b. Kernmantle
c. Braided
d. Braid-on-braid

____ 12. How many strands are generally used to construct laid rope?

a. Two
b. Three
c. Five
d. Seven

____ 13. What type of rope construction is also called twisted construction?

a. Laid
b. Kernmantle
c. Braided
d. Woven

____ 14. Which of the following statements best characterizes braid-on-braid rope?
- a. This construction resists abrasion better than kernmantle.
- b. The construction is similar to two hair braids that are braided together.
- c. All load-bearing strands are exposed and susceptible to abrasion.
- d. The sheath may slide along the inner core of the rope.

____ 15. What rope type is most commonly used as rescue rope?
- a. Dynamic braid-on-braid
- b. Static braid-on-braid
- c. Dynamic kernmantle
- d. Static kernmantle

____ 16. What rope type should be inspected by applying slight tension on the rope and feeling for lumps, depressions, or soft spots?
- a. Kernmantle
- b. Braid-on-braid
- c. Laid
- d. Braided

____ 17. Where should a rope's logbook be kept?
- a. In a durable storage box placed in the same compartment as the rope on the apparatus
- b. In a waterproof envelope placed in a pocket sewn to the side of the rope's storage bag
- c. In a locked file cabinet that only the safety officer can access
- d. In a specified and unlocked cabinet that can be accessed by any of the firefighters who are involved in rope evolutions

____ 18. How should natural fiber ropes be cleaned?
- a. In warm water with a brush
- b. By coiling in a cloth bag and washing in a clothes-washing machine
- c. By wiping or gently brushing
- d. By feeding through a rope washer

____ 19. What clothes-washing machine is recommended for washing synthetic rope?
- a. Top loader with a gentle cycle setting
- b. Top loader with a large capacity tub
- c. Front loader without a plastic window
- d. Front loader with a plastic window

____ 20. Which of the following methods is *not* approved for drying rope?
- a. Air drying
- b. Drying in a hose tower or on a hose rack
- c. Drying in a hose dryer
- d. Looping over a clothesline to dry in the sun

____ 21. When using a clothes washer to clean rope, which of the following wash/rinse combinations is best?

a. Hot/hot
b. Warm/cold
c. Hot/warm
d. Cold/cold

____ 22. What type of cleaning agent should be used to wash rope?

a. Mild soap
b. Detergent
c. Solvent-based cleanser
d. Bleach

____ 23. What is the best storage container and location for rescue ropes?

a. In an airtight storage bag placed in a secured location on apparatus
b. In a plastic bag placed in the apparatus power-tool compartment
c. In a canvas bag placed in a well-ventilated area
d. In a nylon bag placed in the apparatus spare-fuel compartment

____ 24. Which knot is illustrated below?

a. Becket bend or sheet bend
b. Bowline
c. Figure-eight
d. Clove hitch

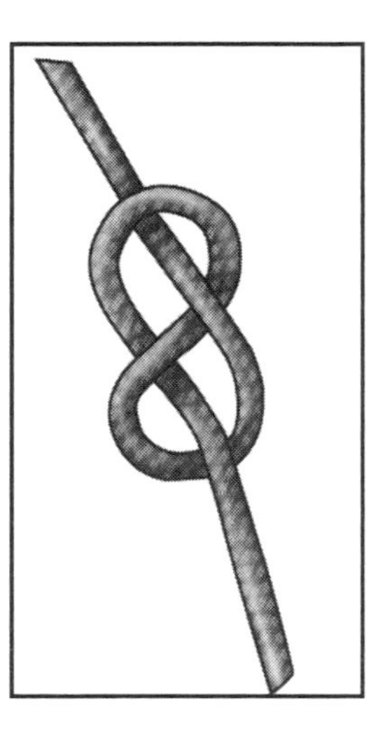

____ 25. Which knot is illustrated below?

a. Becket bend or sheet bend
b. Figure-eight follow through
c. Figure-eight on a bight
d. Clove hitch

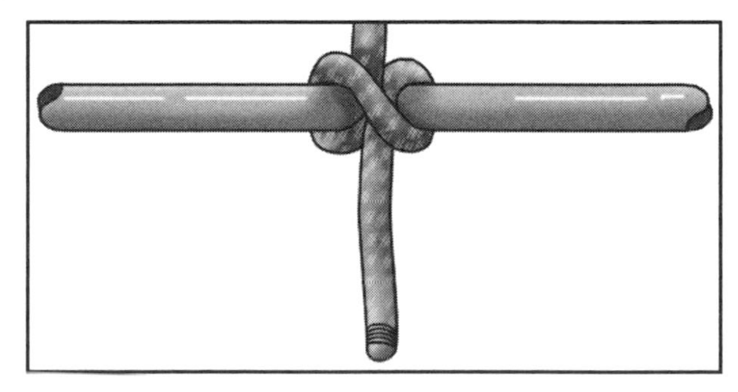

____ 26. Which knot is illustrated below?

a. Bowline
b. Becket bend or sheet bend
c. Figure eight
d. Clove hitch

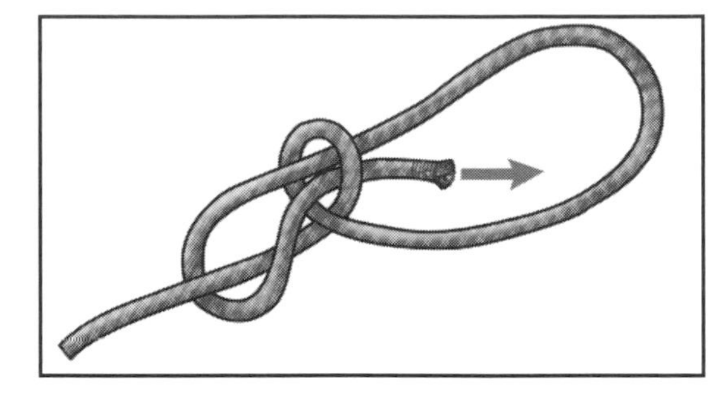

____ 27. Which knot is illustrated below?

a. Figure-eight on a bight
b. Bowline
c. Overhand safety
d. Clove hitch

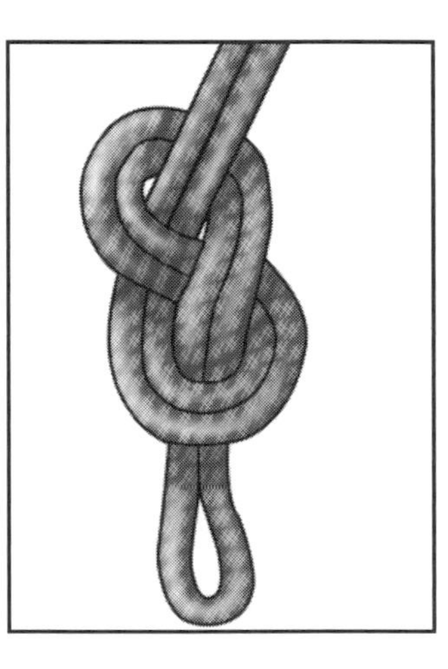

____ 28. Which knot is illustrated below?

a. Figure-eight on a bight
b. Bowline
c. Clove hitch
d. Single overhand

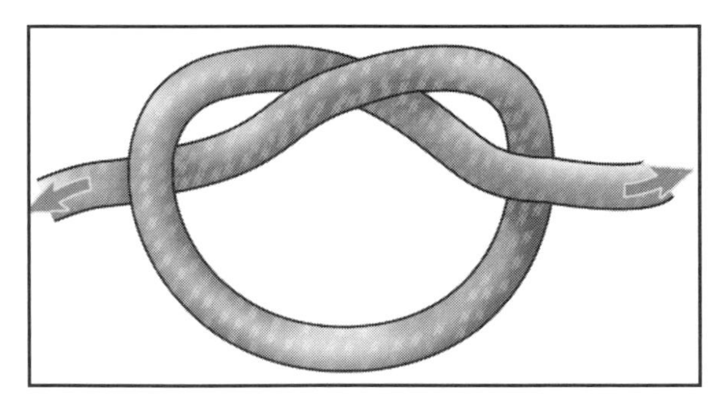

____ 29. What knot is commonly used for safety backup or knot foundation?

a. Overhand
b. Clove hitch
c. Bowline
d. Figure-eight

____ 30. What rope formation is made by simply bending a rope back on itself while keeping the sides parallel?

a. Bight
b. Loop
c. Round turn
d. Overhand

____ 31. What is a good knot for forming a single loop that will not constrict the object around which it is placed?

a. Becket bend or sheet bend
b. Figure-eight
c. Bowline
d. Clove hitch

____ 32. Which knot is most commonly used for attaching ropes to objects and hoisting (with overhand knot)?

a. Sheet bend
b. Clove hitch
c. Figure-eight
d. Figure-eight on a bight

____ 33. What knot is most commonly used for joining ropes of equal diameters?

a. Becket bend or sheet bend
b. Overhand
c. Figure-eight follow through
d. Clove hitch

____ 34. Which knot is most commonly used for securing a loop in the middle of a rope?

a. Figure-eight follow through
b. Clove hitch in the open
c. Figure-eight on a bight
d. Overhand

____ 35. What knot is most commonly used for joining ropes of unequal diameters or joining rope to chain?

a. Becket bend or sheet bend
b. Bowline
c. Figure-eight
d. Figure-eight follow through

____ 36. Which of the following objects should *not* be hoisted?

a. Ladders
b. Fire extinguishers
c. Axes
d. Hoselines

____ 37. What method is best for maintaining control of a rope during a hoisting operation?

a. Underhand
b. Slip-loop
c. Grip-slide
d. Hand-over-hand

____ 38. Which of the following guidelines is recommended for hoisting safety?

a. When working from heights, work in teams to ensure firefighter safety.
b. Work alone when possible to reduce the number of firefighters near the object being hoisted.
c. When working from heights, secure yourself to a team member with a tag line.
d. Do not hoist portable extinguishers or SCBA bottles more than 20 feet *(6 m)* without a tag line.

____ 39. What safety guideline applies during a hoisting operation?

a. Make sure only authorized personnel are in the hoisting area.
b. Be certain that everyone in the hoisting area wears NFPA-approved helmet and safety boots.
c. Ensure that all personnel are clear of the hoisting area.
d. Sound a warning alarm to indicate that personnel in the area should use extra caution.

Identify

E. Identify the missing characteristics of Nylon, Polyester, Polypropylene, and Polyethylene. Write the correct information on the blanks.

Characteristics	Nylon	Polyester	Polypropylene	Polyethylene
Strength	**1.** _______	4*	5*	6*
Wet Strength vs. Dry Strength	85%	100%	**6.** _______	100%
Shock Load Ability	1*	**3.** _______	2*	4*
Floats or Sinks in Water (Specific Gravity)	Sinks (1.14)	Sinks (1.38)	Floats (0.92)	**9.** _______ _______
Elongation at Break (Approximately)	20–34%	15–20%	**7.** _______	10–15%
Melting Point	480°F (249°C)	**4.** _______ _______	330°F (166°C)	275°F (135°C)
Abrasion Resistance	3*	2*	7*	**10.** _______
Resistance:				
Sunlight	**2.** _______	Excellent	Poor	Fair
Rot	Excellent	Excellent	Excellent	Excellent
Acids	Poor	Good	Good	Good
Alkalis	Good	Poor	Good	Good
Oil & Gas	Good	Good	Good	Good
Electrical Conductivity Resistance	Poor	**5.** _______	Good	Good
Storage Requirements	Wet or Dry	Wet or Dry	**8.** _______	Wet or Dry

*** Scale: Best = 1; Poorest = 8**

Source: Wellington Leisure Products, Inc.

F. **Identify the missing characteristics of Manila, Cotton, Kevalr® Aramid, and H. Spectra® Polyethylene. Write the correct information on the blanks**

Characteristics	Manila	Cotton	Kevlar® Aramid	H. Spectra® Polyethylene
Strength	7*	8*	**7.** _______	1*
Wet Strength vs. Dry Strength	115%	**3.** _______	90%	100%
Shock Load Ability	**1.** _______	6*	7*	7*
Floats or Sinks in Water (Specific Gravity)	Sinks (1.38)	Sinks (1.54)	Sinks (1.45)	**9.** _______ _______
Elongation at Break (Approximately)	10–15%	**4.** _______	2–4%	< 4%
Melting Point	Does not melt; chars at 350°F (177°C)	Does not melt; chars at 300°F (149°C)	Does not melt; chars at 800°F (427°C)	**10.** _______ _______
Abrasion Resistance	4*	8*	**8.** _______	1*
Resistance:				
Sunlight	Good	**5.** _______	Good	Good
Rot	Poor	Poor	Excellent	Excellent
Acids	Poor	Poor	Poor	Excellent
Alkalis	Poor	Poor	Good	Excellent
Oil & Gas	Poor	Poor	Good	**11.** _______
Electrical Conductivity Resistance	**2.** _______	Poor	Poor	Good
Storage Requirements	Dry Only	**6.** _______	Wet or Dry	Wet or Dry

*** Scale: Best = 1; Poorest = 8**

Source: Wellington Leisure Products, Inc.

Chapter 7 Rescue and Extrication

Chapter 7 Rescue and Extrication

FIREFIGHTER I

Matching

A. Match to their definitions terms associated with fireground search and rescue. Write the appropriate letters on the blanks.

____ 1. Removal and treatment of victims from situations involving natural elements, structural collapse, elevation differences, etc.

____ 2. Removal and treatment of victims trapped by some type of man-made machinery or equipment

____ 3. Rapid but thorough search performed either before or during fire suppression operations

____ 4. Lying on the back or with the face upward

____ 5. Device such as a stretcher used for carrying a sick or injured person

____ 6. Thorough search performed after initial fire suppression and ventilation operations have been completed

a. Litter
b. Primary search
c. Rescue
d. Supine
e. Extrication
f. Secondary search
g. Perimeter search

True/False

B. Write *True* or *False* before each of the following statements. Correct those statements that are false.

________ 1. Firefighters do not need to conduct a typical search for victims if the fire in a building is small and easily extinguished.

____

____

________ 2. Firefighters do not need to search a burning apartment building if the occupants who have escaped can confirm that it is empty.

____

____

_____ 3. Rescuers should work separately during a primary search.

_____ 4. Primary search personnel should not carry tools with them when they enter a building to conduct a search.

_____ 5. During search operations, negative information is just as important as positive information to ensure a complete search.

_____ 6. During a secondary search, speed is as critical as thoroughness.

_____ 7. During the primary search, doors to rooms not involved in fire should be closed.

_____ 8. To ensure a successful operation, firefighters must set aside concerns for their own safety while searching for victims.

_____ 9. Firefighters should not kick a door to force it open.

_____ 10. With a well-designed incident command or accountability system in place, firefighters can avoid becoming trapped or disoriented within a burning structure.

_______ 11. Firefighters who become trapped should attempt to free themselves or yell for help before activating their PASS devices.

____

____

_______ 12. Firefighters may remove their facepieces to share oxygen with a downed firefighter whose SCBA unit is no longer working.

____

____

_______ 13. Firefighters should not enter a building in which the fire has progressed to the point where viable victims are not likely to be found.

____

____

_______ 14. Often after entering a structure and evaluating the situation, crews must act independently of the operational plan.

____

____

_______ 15. Structures collapse in predictable patterns.

____

____

Multiple Choice

C. Write the letter of the best answer on the blank before each statement.

____ 1. What should firefighters do as soon as they arrive at the scene of a building fire?

a. Immediately prepare for a hose attack.
b. Promptly enter the building and begin a rescue search.
c. Carefully observe the building.
d. Rely solely on size-up information provided by the first arriving officer.

____ 2. When should firefighters identify alternate escape routes from a burning building?

a. During pre-incident planning
b. Before entering the building
c. Upon discovering routes inside the building
d. After determining the need

___ 3. What should firefighters do with information obtained from occupants who have escaped the building?

a. Relay it to the incident commander.
b. Transmit it over the radio to crews entering the building.
c. Forward it to the victims' families and friends.
d. Disregard it, and allow firefighters to confirm the actual status.

___ 4. What are the two objectives of a building search?

a. Finding victims and discovering arson clues
b. Searching for life and searching for fire extension
c. Saving lives and removing bodies
d. Rescuing victims and preventing fire spread

___ 5. Who should conduct the secondary search of a building?

a. Firefighters who conducted the primary search
b. Arson investigators who are also trained in first aid
c. Special search teams assigned to conduct secondary searches at every fire
d. Personnel who were not involved in the primary search

___ 6. How should firefighters ascend and descend stairs in a fire-involved building with reduced visibility?

a. Move up and down stairs slowly; stand upright with arms stretched forward to feel for foreign objects.
b. Move down stairs sideways and in a crouched position, testing each step with one foot before proceeding; move up stairs in the same fashion, testing each step with one hand before proceeding.
c. Move up and down stairs on hands and knees; when ascending, proceed head first, and when descending, proceed feet first.
d. Do not move up and down stairs in situations of reduced visibility.

___ 7. How should rescuers search the fire floor?

a. Begin at the entrance and send one team in a clockwise direction and the other in a counterclockwise direction.
b. Begin at the entrance and search in one direction only.
c. Begin as close to the fire as possible and search back toward the entrance.
d. Begin as far away from the fire as possible and search toward the fire.

___ 8. How should firefighters search each room of a building?

a. Search the perimeter first, and then search the middle of the room.
b. Search the middle of the room, moving in a circular pattern to the perimeter of the room.
c. Search from the front of the room, zigzagging to the back of the room.
d. Search each room differently, giving the most attention to bedrooms and living areas.

____ 9. What should firefighters do when smoke obscures visibility during a primary search?

a. Ventilate the building by opening a window or outside door.

b. Ignore the smoke and continue the rescue operation by feel.

c. Leave the building immediately, and return when visibility is better.

d. Report the condition to the incident commander.

____ 10. What areas are most critical in a multistory building search?

a. Fire floor, the floor directly above it, and the topmost floor

b. Fire floor, the floor directly below it, and the topmost floor

c. Fire floor, the floor directly above it, and the ground-level floor

d. Fire floor, the floor directly below it, and the ground-level floor

____ 11. If there is only one search team, how should the team search a hallway with offices or rooms on each side?

a. One team member should search one side while the other searches the other side.

b. Team members should search down one side and up the other side together.

c. Team members should search from one end of the hallway to the other, crossing from one side of the hallway to the other, searching each room.

d. Team members should call for backup and search only one side of the hallway until more rescuers arrive.

____ 12. Which of the following is *not* a basic rule that firefighters should follow when searching a building?

a. Turn left to leave the room if you turned left to enter the room.

b. Turn right to leave the room if you turned left to enter the room.

c. Exit each room through the same doorway that you entered.

d. Turn the opposite direction used to enter the room when removing a victim.

____ 13. What is the best method for a team to search small rooms?

a. Both team members should move together to search each room.

b. Team members should split up and meet at the entrance at a designated time.

c. One team member should remain at the door while the other member searches the room.

d. One team member should search under beds, in closets, and in all enclosed places while the other member searches open areas.

____ 14. Which of the following is *not* an acceptable method of marking a room that has been searched?

a. Hanging a leather strap on the inside doorknob
b. Drawing an *X* on the door with crayons
c. Using chalk to label a door with an approved mark
d. Drawing a + on a door with masking tape

____ 15. How should firefighters open doors when searching a building?

a. Stand to the side of the door and open it slowly.
b. Stay low and in front of the door and open it quickly.
c. Stay low and to the side of the door and open it slowly.
d. Stand in front of the door and open it quickly.

____ 16. Which of the following actions is recommended for a firefighter who becomes disoriented?

a. Shouting for help periodically
b. Finding a place of relative safety and waiting for help
c. Walking toward a lighted area
d. Following the hoseline in the same direction as the female coupling

____ 17. How can a disoriented firefighter use a window to get help?

a. Lean out of the window and yell for help.
b. Sit on the windowsill with both legs outside, and throw out a helmet, boots, or gloves.
c. Sit on the windowsill with both legs inside the building, and activate the PASS device.
d. Straddle the windowsill and use a flashlight to attract attention.

____ 18. If you are lost or exhausted, what should you do after activating your PASS device?

a. Use a wall to prop yourself into a sitting position.
b. Position your flashlight so that it shines across the floor toward you.
c. Assume either a sitting or prone position in an open area.
d. Lie down near a wall or doorway.

____ 19. Which of the following actions is *not* a recommended safety guideline for search and rescue personnel?

a. Open windows when necessary to dissipate smoke and heat during a search.
b. Maintain contact with a wall when visibility is obscured.
c. Have a charged hoseline at hand whenever possible while working on the fire floor.
d. Report promptly to the supervisor when the search is complete.

____ 20. In what situation would a patient be moved before treatment is provided?

a. The patient is unconscious.
b. The patient requests removal.
c. A nearby area is in danger of fire.
d. The firefighters see no obvious injuries.

____ 21. How should firefighters drag a victim during an emergency move?

a. Holding the feet or ankles
b. Grabbing the clothing in the neck or shoulder area
c. Jackknifing the victim and pulling both the hands and feet
d. Placing the arms above the head and clasping the victim's wrists

____ 22. Which of the following lifts/carries is performed by one rescuer and is best suited to children or small adults?

a. Litter
b. Seat
c. Extremities
d. Cradle-in-arms

____ 23. How many rescuers are required to immobilize a victim on a long backboard when the victim is suspected of having a spinal injury?

a. Two
b. Three
c. Four
d. Five

____ 24. Who directs the actions of rescuers involved in immobilizing a victim suspected of having a spinal injury?

a. Rescuer who applies and maintains in-line stabilization
b. Medical personnel via radio communication
c. One of the rescuers moving the victim from the ground to the backboard
d. Rescuer with the highest rank and most experience

____ 25. What kind of chair would be most suitable for a chair lift/carry?

a. Folding chair
b. Kitchen table chair
c. Reclining lawn furniture
d. Futon chair

____ 26. Who should attempt to rescue heavily trapped victims of a building collapse?

a. Rescue workers who have EMT training or advanced CPR training
b. Specially trained rescue workers who have knowledge of building construction and collapse
c. The most experienced first arriving rescue workers
d. Any firefighter who has undergone Firefighter I and Firefighter II training and has been certified

Identify

D. Identify the following abbreviations associated with rescue. Write the correct interpretation before each.

_______________________________________ 1. IC

_______________________________________ 2. CPR

FIREFIGHTER II

Matching

A. Match to their definitions terms associated with rescue and extrication. Write the appropriate letters on the blanks.

____ 1. Step-up transformer that converts a vehicle's DC current into an AC current

____ 2. To line or support with a framework of timber

____ 3. Term for dressed and bandaged wounds and splinted fractures

____ 4. Two sheets of glass bonded to a sheet of plastic sandwiched between them

____ 5. Any of a variety of means by which unstable structures or parts of structures can be stabilized

____ 6. Removing small rubble and debris to create a path to a victim whose location is known

____ 7. Searching for a victim who has been submerged in water for such a long period of time that survival is unlikely

____ 8. Wooden or metal frame containing one or more pulleys

____ 9. Assembly of ropes and blocks through which a line passes to multiply pulling force

a. Tunneling
b. Block
c. Packaging
d. Crib
e. Inverter
f. Recovery
g. Tackle
h. Tempered glass
i. Shoring
j. Laminated glass

True/False

B. Write *True* or *False* before each of the following statements. Correct those statements that are false.

_______ 1. Inverters are the most common power source used for emergency services.

____

____

_______ 2. Mounted generators with a separate engine are noisy and make communication near them difficult.

____

____

_______ 3. Equipment should not be interchanged between mutual aid departments if adapters must be used to do so.

____

____

_______ 4. Hydraulic shears can cut almost any metal object that fits between its blades.

____

____

_______ 5. A combination hydraulic spreader/shears has greater spreading and cutting capabilities than each individual unit has.

____

____

_______ 6. Bar screw jacks are excellent for supporting collapsed structural members.

____

____

_______ 7. When working under a load supported entirely by a jack, another team member should be available to monitor the stability of the load.

____

____

_______ 8. Rescuers should not use oxygen-supply tanks to power pneumatic tools.

_______ 9. Sparks produced while cutting metal with pneumatic chisels may provide an ignition source for flammable vapors.

_______ 10. Air bags for lifting or displacing objects can be positioned against any surface.

_______ 11. Block and tackle systems should be used to stabilize or lift victims.

_______ 12. Ideally, one rescuer per vehicle would be assigned to assess the scene of a vehicle accident.

_______ 13. The rescuer(s) who check(s) the vehicles should also be assigned to subsequently survey the entire area around the scene.

_______ 14. Before attempting to stabilize a vehicle, rescuers should first test its stability in the position it is found.

_______ 15. Rescuers should avoid placing any part of their bodies under a vehicle while placing stabilizing devices.

_______ 16. After disconnecting power to an electrically operated air bag restraint system, sufficient power to deploy an air bag can remain in the system from 10 seconds to 10 minutes.

____

____

_______ 17. At the scene of an automobile accident, rescuers should remove the vehicle from around the patient and not the reverse.

____

____

_______ 18. When it is necessary to break a window to gain primary access to the victim of a motor vehicle accident, the rescuer should choose a window near the victim.

____

____

_______ 19. Regardless of vehicle construction material or type, rescuers should remove roofs by cutting the front posts and folding the roof back on itself.

____

____

_______ 20. Downed electrical lines can create energized fields that extend several feet *(meters)* from the point of contact.

____

____

_______ 21. The thickness of ice directly correlates to its strength.

____

____

_______ 22. Rescuers with electrical equipment training may adjust the mechanical system of an elevator instead of calling an elevator mechanic to do so.

____

____

_________ 23. Rescuers should minimize conversation with passengers who are trapped in an elevator.

____

____

_________ 24. Escalators should be stopped before firefighters advance hoselines up or down them.

____

____

_________ 25. Shoring is used to both stabilize and move heavy objects.

____

____

Multiple Choice

C. Write the letter of the best answer on the blank before each statement.

____ 1. Which description best characterizes a portable generator?

a. Is powered by hydraulic or power take-off systems
b. Has fixed flood lights wired directly to it through a switch
c. Makes more noise than a mounted generator
d. Is light enough to be carried by two people

____ 2. What is the normal wattage for portable lights?

a. 250 to 750
b. 300 to 1,000
c. 350 to 1,200
d. 400 to 1,500

____ 3. Which of the following methods is *not* one of the recommended ways to store electrical cord?

a. Figure-eight winding
b. Coiling
c. Using portable cord reels
d. Using fixed, automatic rewind reels

____ 4. Which NFPA standard requires all outlets in a junction box to be equipped with ground-fault circuit interrupters?

a. 70E
b. 90B
c. 1972
d. 1970

____ 5. What is true about manual hydraulic tools?
- a. They receive power through a hose reel line.
- b. They operate by a pump lever.
- c. They have superior power to powered hydraulic tools.
- d. They operate on compressed air.

____ 6. Up to how many inches *(millimeters)* of space and how much psi *(kPa)* can powered hydraulic spreaders produce?
- a. 7 *(180)* and 30,000 *(206 850)*
- b. 32 *(813)* and 22,000 *(154 000)*
- c. 51 *(1 295)* and 18,000 *(124 110)*
- d. 63 *(1 600)* and 15,000 *(104 000)*

____ 7. What is the primary advantage of the porta-power tool system?
- a. Designed for heavy lifting applications
- b. Excellent device for shoring or stabilizing
- c. Convenient for operating in narrow places
- d. Time-efficient because of abundance of accessories

____ 8. What is the lifting capacity of most hydraulic jacks?
- a. 20 tons *(20.3 t)*
- b. 30 tons *(30.5 t)*
- c. 40 tons *(40.6 t)*
- d. 50 tons *(50.8 t)*

____ 9. What is the primary use of a bar screw jack?
- a. Moving an object
- b. Lifting an object
- c. Stabilizing an object
- d. Cribbing during rescue operations

____ 10. Which jacks are the least stable?
- a. Bar screw
- b. Trench screw
- c. Hydraulic
- d. Ratchet

____ 11. What is a recommended procedure for handling the cribbing used in rescue operations?
- a. Paint entire length of cribbing for easy size determination.
- b. Varnish entire length, but paint ends only.
- c. Stack in a storage compartment with the grab handles facing inward.
- d. Store on end inside a storage crate.

____ 12. What tools use compressed air for power?
- a. Electric
- b. Hydraulic
- c. Electronic
- d. Pneumatic

____ 13. Which of the following tasks is *not* typically accomplished by an air chisel, operating between 100 and 150 psi *(700 kPa* and *1 050 kPa)*?

a. Cutting medium-gauge sheet metal
b. Breaking locks and driving in plugs
c. Driving nails into masonry
d. Popping rivets and bolts

____ 14. What tool creates an anchor point above a utility cover or other opening and allows rescuers to be safely lowered into and out of confined spaces?

a. Winch
b. Tripod
c. Come-along
d. Block and tackle system

____ 15. What cord material is most commonly used in vehicle-mounted winches?

a. Chain or steel cable
b. Common or hardware chain
c. Nylon or Kevlar® cable
d. Proof coil chain

____ 16. What is the safest distance for a winch operator to remain from the winch?

a. No less than 10 feet *(3 m)* from the load
b. At least 15 feet *(5 m)* from the winch
c. About 1 foot *(0.3 m)* for every pound *(0.5 kg)* of approximate load weight
d. Farther than the distance from the winch to the load

____ 17. What is the maximum temperature allowed for any object that comes into contact with a lifting bag?

a. 220°F *(104°C)*
b. 250°F *(121°C)*
c. 280°F *(138°C)*
d. 310°F *(154°C)*

____ 18. Which of the following rules should be followed when stacking lifting bags during an emergency operation?

a. Never stack bags more than three high.
b. Always inflate the top bag first.
c. When possible, choose a multicell bag rather than multiple bags.
d. For additional support, wedge smaller bags below larger bags.

____ 19. What kind of building collapse pattern is depicted below?

a. V-shaped collapse
b. Pancake collapse
c. Cantilever collapse
d. Lean-to collapse

FIREFIGHTER II

____ 20. What kind of building collapse pattern is depicted below?

a. V-shaped collapse
b. Pancake collapse
c. Cantilever collapse
d. Lean-to collapse

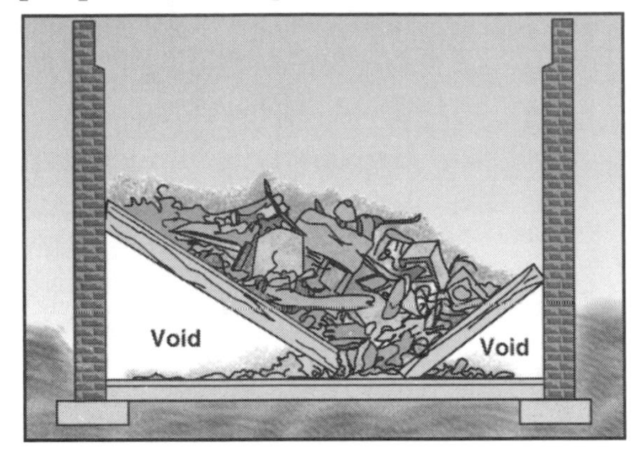

____ 21. What kind of building collapse pattern is depicted below?

a. V-shaped collapse
b. Pancake collapse
c. Cantilever collapse
d. Lean-to collapse

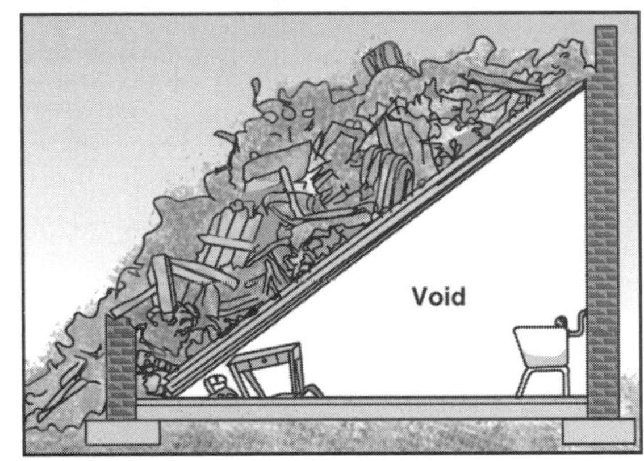

____ 22. What kind of building collapse pattern is depicted below?

a. V-shaped collapse
b. Pancake collapse
c. Cantilever collapse
d. Lean-to collapse

____ 23. Which of the following is an example of an actual or potential environmental hazard at a structural collapse scene?

a. Damaged utilities
b. Unstable debris
c. Exposed wiring and rebar
d. Irregularly shaped rubble

____ 24. What should emergency personnel do first upon arriving at a vehicle accident scene?

a. Begin extrication efforts.

b. Cordon off the area.

c. Reassure victims and bystanders.

d. Start size-up.

____ 25. How should firefighters park apparatus at the scene of a motor vehicle accident?

a. In a circle completely encompassing the accident scene

b. On the side of the accident farthest from the traffic

c. Adjacent to the accident and protecting the scene from oncoming traffic

d. Across the roadway from the scene of an accident and out of the way of oncoming traffic

____ 26. What is the safest way to prevent the horizontal motion of a vehicle involved in an accident?

a. Chock the wheels.

b. Set the emergency brake.

c. Put the automatic transmission in park or the manual transmission in gear.

d. Abut emergency apparatus at both ends of the vehicle.

____ 27. Approximately how fast do SRS and SIPS deploy?

a. 100 mph *(161 kmph)*

b. 200 mph *(322 kmph)*

c. 300 mph *(483 kmph)*

d. 400 mph *(644 kmph)*

____ 28. How can firefighters prevent mechanically operated vehicle restraint systems from deploying?

a. Avoid contact with the air bag during extrication operations.

b. Use the key-operated switch to disable the air bag and wait for the reserve power to drain.

c. Disconnect both battery cables and wait for the reserve power to drain.

d. Sever the connection between the sensor and the air bag inflation unit.

____ 29. Which of the following procedures is recommended for removing a windshield?

a. Instruct the rescuer and passengers inside the vehicle to hold the tarp or protective blanket over themselves.

b. Use three rescuers for total windshield removal: one on each side and one on the front of the hood to cut along the bottom edge.

c. Hold a backboard over people inside the vehicle for added protection against loose glass and tools.

d. Remove or partially lay back the roof before starting on the windshield.

____ 30. What is the recommended tool for removing tempered glass?

a. Claw hammer
b. Pneumatic hammer
c. Hatchet
d. Phillips™ screwdriver

____ 31. Where is the B-post located on a hatchback vehicle?

a. Where the front door is connected to the body
b. Where the hatchback door is connected to the body
c. At the door handle end post
d. Nowhere; three-door cars do not have B-posts

____ 32. When should firefighters use the tunneling process to reach victims?

a. Immediately if the victim is buried under tons *(tonnes)* of rubble
b. At the same time that other rescuers are attempting to shore up the area
c. After making verbal contact with the victim or hearing noise from under the rubble
d. As a last resort after all other means of reaching the victim have failed

____ 33. What determines whether or not personnel should be sent into a trench for a rescue operation?

a. Safety of the rescuer
b. Age of victim(s)
c. Duration of entrapment
d. Number of victims trapped

____ 34. When ladders are placed in trenches for trench rescue, how far above the top of the trench should the ladder extend?

a. 1 foot *(0.3 m)*
b. 2 feet *(0.6 m)*
c. 3 feet *(1.0 m)*
d. 4 feet *(1.2 m)*

____ 35. If fire and rescue personnel are not specially trained to perform rescues in caves and tunnels, what should they do at these emergency scenes?

a. Team with trained rescue personnel.
b. Confine their activities to aboveground support.
c. Lead backup crews into the tunnel or cave after trained crews have finished their shifts.
d. Exit the scene to allow trained rescue personnel to take over the operation.

____ 36. When responding to a rescue call involving electricity, which of the following actions should rescuers take?

a. Call for the power provider to respond.
b. Cut the electrical wires that are causing the hazard.
c. Assume that the electrical lines or equipment are no longer energized.
d. Stay outside a 3-foot *(1 m)* perimeter around downed wires.

____ 37. Which water rescue method requires special training to perform?

a. Stretch
b. Reach
c. Row
d. Throw

____ 38. Which ice rescue method requires special training to perform?

a. Pull
b. Go
c. Reach
d. Stretch

____ 39. When should firefighters identify possible outside sources for extricating victims entangled in industrial machinery?

a. During pre-incident planning
b. While performing the initial survey of an accident
c. After consulting plant maintenance personnel
d. Before beginning any extrication attempt

____ 40. When should firefighters have an elevator mechanic dispatched to the scene of an elevator rescue?

a. Upon arrival at the scene
b. When a building has more than three stories
c. If the elevator appears to have a computer problem
d. After rescuers' attempts to mechanically adjust the elevator have failed

Identify

D. Identify the following abbreviations associated with extrication. Write the correct interpretation before each.

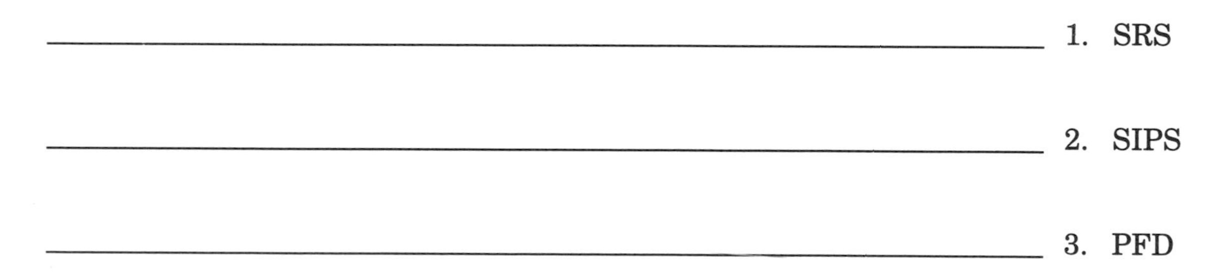

____________________ 1. SRS

____________________ 2. SIPS

____________________ 3. PFD

E. **Identify the following rescue and extrication tools. Write the correct name below each picture.**

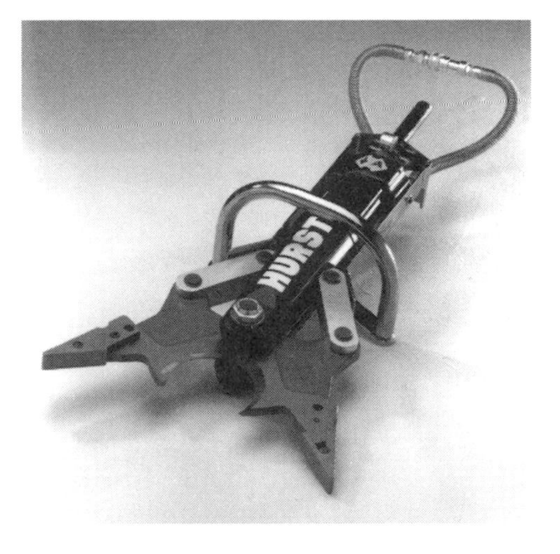

________________________________ 1.

________________________________ 2.

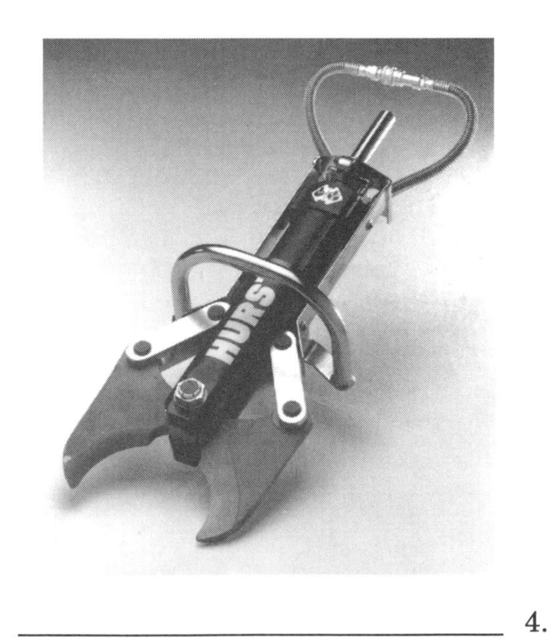

________________________________ 3.

________________________________ 4.

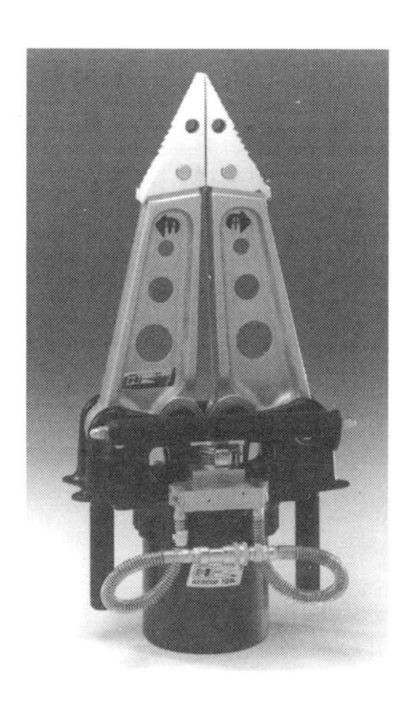

_______________ 5.

_______________ 6.

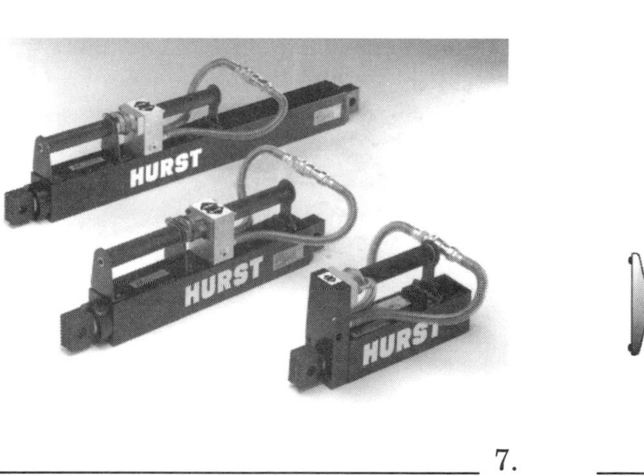

_______________ 7.

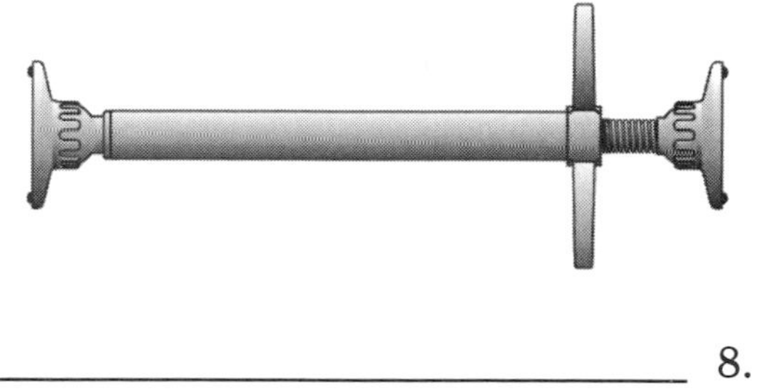

_______________ 8.

Photos for Questions 2, 4, 5, and 7 courtesy of Hale Fire Pump Co, Inc.

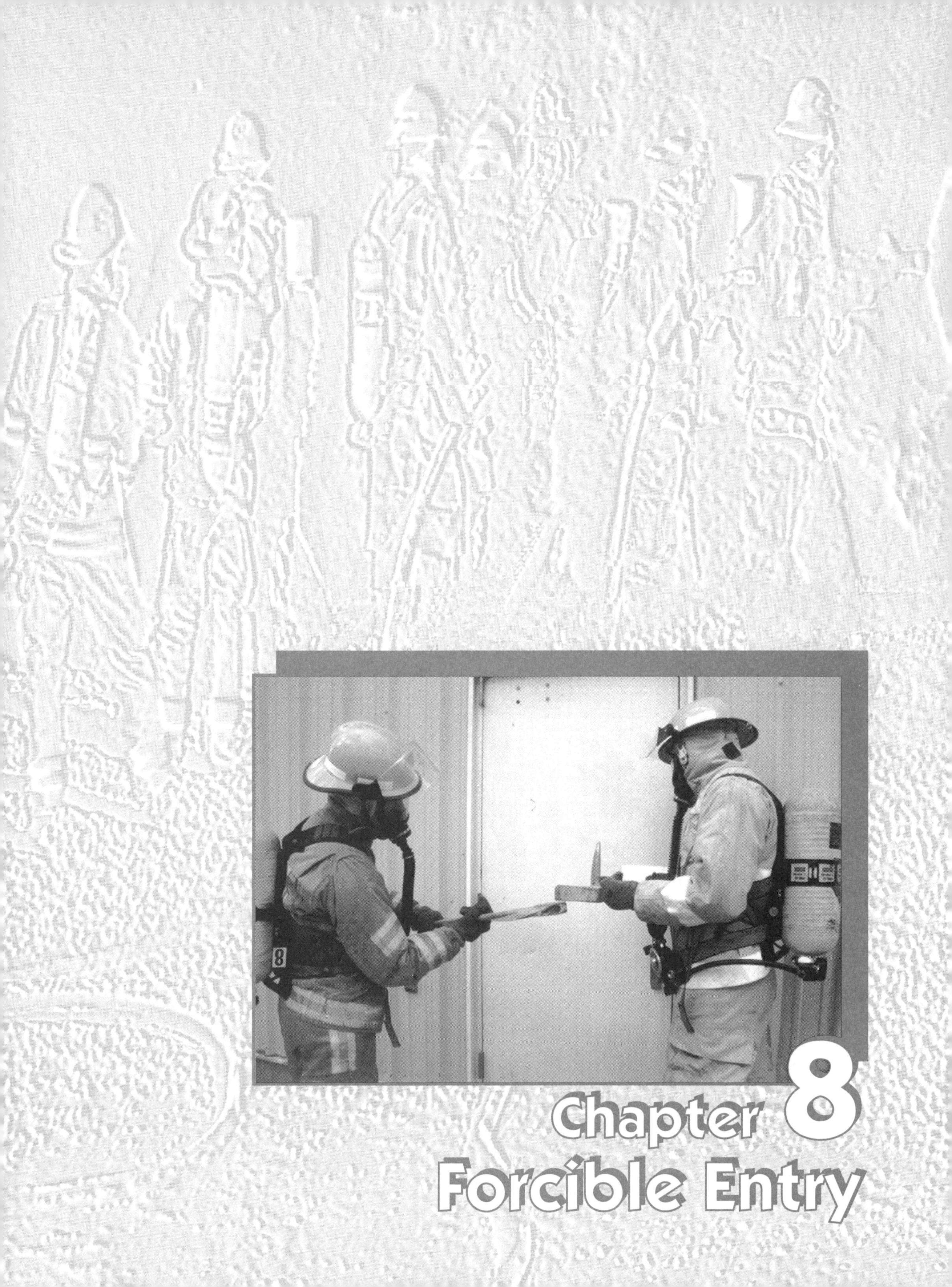

Chapter 8
Forcible Entry

Chapter 8 Forcible Entry

FIREFIGHTER I

FIREFIGHTER I

Matching

A. Match to their definitions terms associated with forcible entry. Write the appropriate letters on the blanks.

____ 1. Gaining access to a structure whose normal means of access is locked, blocked, or nonexistent

____ 2. Cutting tool with a thin arched blade set at right angles to the handle

____ 3. Rotating or sliding piece in a mechanical linkage used in transforming rotary motion into linear motion or vice versa

____ 4. Support about which a lever turns

____ 5. Makeshift extension added to a prying tool to lengthen the handle and provide additional leverage

____ 6. Opening a hole in a wall to gain access

a. Cam
b. Egress
c. Breaching
d. Adze
e. Fulcrum
f. Forcible entry
g. Cheater

True/False

B. Write *True* or *False* before each of the following statements. Correct those statements that are false.

________ 1. Firefighters acquire and master forcible entry skills during on-the-job training.

____

____

________ 2. Selecting the appropriate tool is a primary concern during forcible entry size-up.

____

____

_______ 3. Only power saws can efficiently cut any material.

_______ 4. Firefighters use handsaws for forcible entry procedures requiring speed.

_______ 5. If necessary, firefighters should use power saws to gain access to a fire-involved fertilizer manufacturing plant.

_______ 6. Firefighters should not use cutting torches unless they have undergone training and considerable practice.

_______ 7. Most prying tools can also be used effectively as striking tools.

_______ 8. Circular saw blades that look alike are not necessarily interchangeable.

_______ 9. A dull saw is more likely than a sharp one to cause an accident.

_______ 10. When entering a building with a pike pole, a firefighter should carry the tool with the tool head down, close to the ground and ahead of the body.

_________ 11. Firefighters should use extreme caution when carrying running power tools from one area to another.

____

____

_________ 12. The entire inside of a hollow core door is hollow.

____

____

_________ 13. Metal swinging doors set in metal doorjambs have very little "spring."

____

____

_________ 14. Firefighters should force a revolving door rather than a swinging door when faced with a choice between the two.

____

____

_________ 15. Firefighters should not block open overhead doors.

____

____

_________ 16. If at all possible, firefighters should avoid trying to force entry through an overhead rolling door.

____

____

_________ 17. Most interior fire doors do not lock when they close.

____

____

_________ 18. Firefighters should wear goggles when breaking glass to enter a fire building.

____

____

_______ 19. If a door and lock are suitable for through-the-lock forcible entry, firefighters should use this method rather than conventional forcible entry methods.

_______ 20. When working alone, a firefighter should use a power saw or cutting torch to cut a loose padlock.

_______ 21. Windows are the preferred entry point into a fire building.

_______ 22. Safety glass is approximately 30 times stronger than Lexan®.

_______ 23. Firefighters can use a battering ram to break masonry walls.

_______ 24. If no studs can be located in a metal wall, firefighters should assume that the wall bears the entire load of the structure.

_______ 25. A wood floor in a building ensures that firefighters will be able to easily breach the floor if it becomes necessary.

_______ 26. The best power supply sources for electric saws are those found at the scene, but a portable generator can be used if necessary.

_______ 27. Firefighters should take time to remove carpet or roll it to one side before cutting a floor.

____

____

Multiple Choice

C. Write the letter of the best answer on the blank before each statement.

____ 1. How much do typical pick-axe heads weigh?

a. 3 or 5 pounds *(1.4 kg or 2.3 kg)*
b. 4 or 6 pounds *(1.8 kg or 3 kg)*
c. 5 or 7 pounds *(2.3 kg or 3.2 kg)*
d. 6 or 8 pounds *(3 kg or 3.6 kg)*

____ 2. Which of the following saws is *not* a handsaw?

a. Coping
b. Reciprocating
c. Carpenter's
d. Keyhole

____ 3. What saw blades are less prone to dulling than standard blades?

a. Large-toothed
b. Fine-toothed
c. Carbide-tipped
d. Lexan®-rated

____ 4. What power saw has a short, straight blade that moves forward and backward with an action similar to that of a handsaw?

a. Reciprocating
b. Ventilation
c. Carpenter's
d. Coping

____ 5. What saw should not be used to cut metal?

a. Carpenter's
b. Ventilation
c. Rotary
d. Hacksaw

____ 6. Which of the following security devices can be cut with bolt cutters?

a. High-security chains
b. Case-hardened hasps
c. Ferrous alloy padlock shackles
d. Iron bars

____ 7. What is an example of a manual prying tool?

a. Kelly tool
b. Chisel
c. Mallet
d. Maul

____ 8. What rescue tool also has forcible entry uses?
   a. Bar screw jack
   b. Pneumatic hammer
   c. Hydraulic ram
   d. Block and tackle

____ 9. What tool is extremely valuable when more than one door must be forced such as in apartments or hotels?
   a. Hydraulic door opener
   b. Hydraulic rescue spreader tool
   c. Hydraulic ram
   d. San Francisco hook

____ 10. What is the best tool for breaking a window from a distance to stay out of the way of falling glass?
   a. Battering ram
   b. Pike pole
   c. Maul
   d. Punch

____ 11. What hook can be used for leverage?
   a. Clemens
   b. Plaster
   c. Multipurpose
   d. Roofman's

____ 12. When is the best time for a firefighter to determine the appropriate tools for forcible entry into a building?
   a. En route to the fire
   b. As they are needed
   c. After primary size-up
   d. During pre-incident planning

____ 13. What is the recommended method for a firefighter to learn to properly use and maintain forcible entry tools?
   a. Read manufacturer instructions and learn SOPs for tool safety.
   b. Ask an experienced firefighter about the tools and watch others use them.
   c. Learn departmental SOPs for tool safety and watch others use the tools.
   d. Follow posted guidelines on tool safety and ask an experienced firefighter about the tools.

____ 14. What should a firefighter do if additional leverage is needed for a prying task?
   a. Add a piece of pipe to the end of the tool.
   b. Lay another tool perpendicular to the prying tool and add pressure.
   c. Stop using the tool and choose another one.
   d. Wedge the prying tool and use a striking tool to increase leverage.

____ 15. Where should firefighters store the extra blades for circular saws?
   a. With the power tools and spare fuel
   b. In a storage unit free of hydrocarbons
   c. In a moderately humid atmosphere
   d. In a crate filled with sawdust

____ 16. What type tools should be carried with the head close to the ground?
a. Leverage
b. Power
c. Prying
d. Striking

____ 17. What is a proper procedure for maintaining a tool's wooden handle?
a. Apply a coat of boiled linseed oil.
b. Soak in warm water with a mild detergent.
c. Paint or varnish the wood.
d. Wrap with electrical tape when splintered.

____ 18. How should firefighters care for unprotected metal surfaces on tools?
a. File off spurs, burrs, or sharp edges.
b. Coat with a metal protectant containing 1-1-1-trichloroethane.
c. Oil the surface with a medium-to-heavy machine oil.
d. Paint the unprotected metal surface.

____ 19. Which of the following statements is true about axe heads?
a. Sufficiently sharpening thick-bodied blades enables them to cut, rather than bounce off, ordinary surfaces.
b. Adequately sharpening extremely thin-bodied blades prevents them from breaking on rough surfaces such as gravel roofs.
c. Painting axe heads can cause the axe to stick and bind to cutting surfaces during forcible entry procedures.
d. Painting axe heads protects the metal surface from corrosive chemical agents encountered during a fire.

____ 20. What is the *first* step a firefighter should take in forcible entry of doors and windows?
a. Identify the type of locking mechanism.
b. Determine the proper tool for the job.
c. Identify the construction type.
d. Try to open the door or window.

____ 21. Which of the following statements is *always* true?
a. Access doors to residences swing inward.
b. Commercial building doors swing outward.
c. Industrial building doors swing outward.
d. If you can see a door's hinges, it swings toward you.

____ 22. Which of the following guidelines should *not* be followed when forcing a door?
a. If the door does not force using the technique chosen, choose another.
b. If the door proves to be too well secured, call for more help.
c. If the tool chosen is inadequate, choose another.
d. If too much time is spent forcing a door, the task is counterproductive.

_____ 23. Which of the following types is *not* a wooden swinging door category?
- a. Core
- b. Panel
- c. Slab
- d. Ledge

_____ 24. What doorjamb can be easily removed with prying tools?
- a. Rabbeted
- b. Recessed
- c. Stopped
- d. Opened

_____ 25. Which of the following doors is made of solid wood members inset with glass, wood, or plastic frames?
- a. Slab
- b. Solid core
- c. Ledge
- d. Panel

_____ 26. What statement best describes all solid core doors?
- a. Made of thick planks that have been tongue and grooved together
- b. Have dense centers with plywood veneer coverings
- c. Contain compressed mineral material for fire resistance
- d. Filled with material used for insulation or soundproofing

_____ 27. What type doors are most often found on warehouses, storerooms, barns, and sheds?
- a. Batten
- b. Slab
- c. Panel
- d. Solid core

_____ 28. Which of the following doors would generally be considered impractical to force?
- a. Hollow core doors with wooden frames set in masonry
- b. Solid core doors with wooden frames set in heavy timber
- c. Metal doors with metal frames set in masonry
- d. Panel doors with metal frames set in heavy timber

_____ 29. When forcing a metal door, what tool will likely work best?
- a. Battering ram
- b. Sledgehammer
- c. Hydraulic tool
- d. Keyhole saw

_____ 30. How do burglar blocks on patio sliding doors affect forcible entry?
- a. They make entry nearly impossible without excessive damage.
- b. They are difficult to see during size-up.
- c. They do not complicate forcible entry procedures.
- d. They render the door totally impenetrable.

_____ 31. Which of the following is *not* a type of revolving door collapse mechanism?
- a. Panic-proof
- b. Drop-arm
- c. Metal-braced
- d. Alarm-indicated

____ 32. How are overhead doors classified for forcible entry?

a. Small, medium, and large
b. Sectional, rolling steel, and slab
c. Wood, metal, and fiberglass
d. Manual, motor driven, and remotely controlled

____ 33. How should firefighters access a rolling steel door with a rescue saw or cutting torch?

a. Cut a small square hole next to the locking mechanism and reach through the hole to unlock the door.
b. Cut a large rectangular hole approximately the size of a standard doorway for firefighter entry and egress.
c. Cut a triangular opening through which firefighters can crawl.
d. Cut a small opening on each side of the door and use a crowbar to access the manual release system located near the roller track.

____ 34. What door assembly is placed in a wall opening rated as a fire-barrier wall?

a. Panel door with Lexan® glass
b. Ledge door with a self-closing mechanism
c. Fire-door assembly
d. Tubular metal with tempered glass

____ 35. What fire door type returns each time to the closed position after it is opened?

a. Automatic-closing
b. Self-closing
c. Self-propelling
d. Power-driven

____ 36. What precautionary measure should firefighters take when passing through an exterior door?

a. Make certain the door closes behind them.
b. Deactivate the automatic-closing device.
c. Identify the door with standard rescue marking.
d. Block the door open.

____ 37. Where are counterbalanced fire doors generally found?

a. Class A openings
b. Hospital stairwells
c. Freight elevators
d. Passenger elevators

____ 38. What lock is a surface-mounted, add-on lock?

a. Mortise
b. Rim
c. Bored
d. Surface

____ 39. In what category is the *key-in-knob* lock?

a. Mortise
b. Rim
c. Bored
d. Surface

____ 40. What characterizes a *regular* padlock?

a. Shackles ½ inch *(13 mm)* or less in diameter
b. Shackles ¼ inch *(6 mm)* or less in diameter
c. Toe and heel locking
d. Case-hardened shackles

____ 41. Which of the following statements is true regarding the rapid-entry key box system?

a. Only the fire department has a key to fit all of the boxes in the jurisdiction.
b. Proper mounting is the responsibility of the fire department in whose jurisdiction the box is located.
c. The property owner places all of the keys inside the box and locks the box with a master key.
d. The box must be mounted on an interior wall in a high-visibility location.

____ 42. For what type door is the goal of forcible entry to open the space between the door and the doorjamb to allow the bolt to slip from its keeper?

a. Inward swinging
b. Outward swinging
c. Double swinging
d. Revolving

____ 43. What kind of door is most often secured by a drop-bar assembly?

a. Inward swinging
b. Outward swinging
c. Double swinging
d. Revolving

____ 44. Which is a standard procedure for forcing a drop-bar assembly?

a. Use a crowbar to lift the bar up and out of its stirrup.
b. Use a battering ram to break the bar from its stirrup.
c. Insert a carpenter's saw or a coping saw between the doors and cut the bar.
d. Cut a hole in the door just below the bar and reach in to release the bar.

____ 45. What is a characteristic of plate glass?

a. Resists heat
b. Shatters into thousands of cubelike pieces
c. Difficult to break
d. Cracks and breaks into shards

____ 46. What is the preferred method of forcible entry for many commercial doors, residential security locks, and high-security doors?

a. Conventional
b. Through-the-lock
c. Fork end pry
d. Adze end jimmy

____ 47. Which of the following tools causes the most damage to a door during forcible entry?

a. A-tool
b. K-tool
c. J-tool
d. S-tool

____ 48. What tool was developed as a result of manufacturers putting collars or protective cone-shaped covers over locks to prevent anyone from using lock-pulling devices on them?

a. A-tool
b. K-tool
c. J-tool
d. S-tool

____ 49. What tool was designed to fit through the space between double swinging doors equipped with panic hardware?

a. A-tool
b. K-tool
c. J-tool
d. S-tool

____ 50. Which of the following tools is used for forcible entry of case-hardened padlocks?

a. Flat-head axe
b. Duck-billed lock breaker
c. Halligan-type bar
d. Bolt cutters

____ 51. What padlocks must first be stabilized with locking pliers and chain by one firefighter and then cut with a power saw or torch by a second firefighter?

a. Standard
b. Steel-cased
c. Case-hardened
d. Loose

____ 52. Where should firefighters cut a wire fence?

a. Near the ground in a triangular shape to minimize damage to the fence
b. Halfway between two posts to provide adequate access
c. Near posts to lessen the danger of injury from the whip coil
d. Away from the gate to decrease tension on the fencing

____ 53. What tool works best for breaking wire glass?

a. Axe pick
b. Sledgehammer
c. Power saw
d. Kelly tool

____ 54. How do metal windows differ from wooden windows in regard to forcible entry?

a. There are no differences between the two.
b. Metal windows are easier to pry.
c. Metal window-locking mechanisms pull out of sashes easier.
d. Metal window-locking mechanisms tend to jam.

____ 55. With what tool can firefighters safely remove glass from a window?

a. An 8- or 10-foot *(2.4 m or 3 m)* pike pole or hook
b. A 3- to 4-foot *(1 m to 1.2 m)* crowbar
c. A ball peen hammer with a pipe extension
d. Any 3- to 4-foot *(1 m to 1.2 m)* piece of wood or metal

____ 56. What is true about most casement windows?

a. Are also known as awning or jalousie windows
b. Consist of two sashes mounted on side hinges that swing outward
c. Are commonly wide windows with two locks and one cranking device
d. Can be accessed only by applying pressure to the push bar

____ 57. Which of the following actions is a recommended procedure for forcing a hinged window?

a. Break the top pane and reach in and down to unlock the latch.
b. Force or cut the screen in the upper right-hand corner.
c. Break the lowest pane and reach in and up to unlock the latch.
d. Remove the screen completely before breaking the glass.

____ 58. Where should a firefighter break the glass to gain entry through a projected window?

a. At the top of the uppermost pane
b. Directly opposite the latch
c. At the top of the bottom pane
d. As close to the latch as possible

____ 59. What windows are the most difficult to force?

a. Jalousie
b. Hinged
c. Double-hung
d. Factory

____ 60. Which of the following statements most accurately describes the effect of forcible entry tools on Lexan®?

a. Large-toothed rotary power saw blades will melt the surface and cause the blade to bind.
b. Intense cold followed by a sharp blow will shatter the material.
c. Carbide-tipped rotary power saw blades will skid off the surface.
d. Conventional forcible entry tools work better on this material than on acrylic or safety glass.

____ 61. Which of the following wall breaching tools is *not* time efficient?

a. Hydraulic spreaders
b. Rotary rescue saws
c. Air chisels
d. Battering rams

____ 62. How wide should firefighters open a gypsum wallboard or plaster wall during forcible entry?

a. Large enough to use as the main point of entry and egress
b. At least three bays wide
c. As wide as possible without removing a stud
d. Up to 3 × 3 feet *(1 m by 1 m)*

____ 63. What is the best tool for forcing entry through a metal wall?

a. Rotary or rescue saw
b. Oxyacetylene torch
c. Metal-cutting power saw
d. Pneumatic hammer

____ 64. What shape of entry hole most evenly distributes a load-bearing metal wall's load, thus reducing the risk of collapse?

a. Square
b. Rectangle
c. Triangle
d. Circle

____ 65. At what maximum distance are the joists of a wood floor usually spaced?

a. 14 inches *(350 mm)*
b. 16 inches *(400 mm)*
c. 18 inches *(450 mm)*
d. 20 inches *(500 mm)*

____ 66. For which of the following applications would a puncture or penetrating nozzle be most effective?

a. Lath and plaster partitions
b. Fire doors
c. Concrete floors
d. Steel exterior walls

Identify

D. Identify the following cutting tools used for forcible entry. Write the correct name below each.

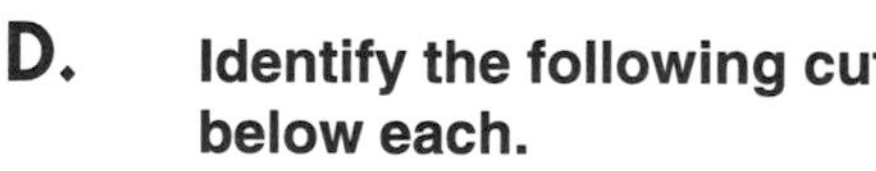

________________ 1.

________________ 2.

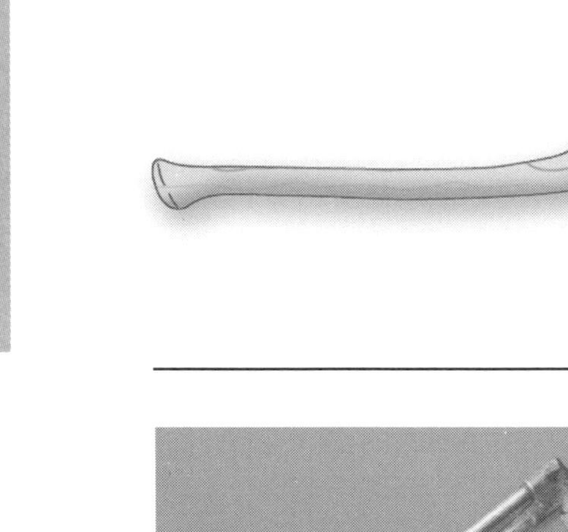

______________________ 3.

______________________ 4.

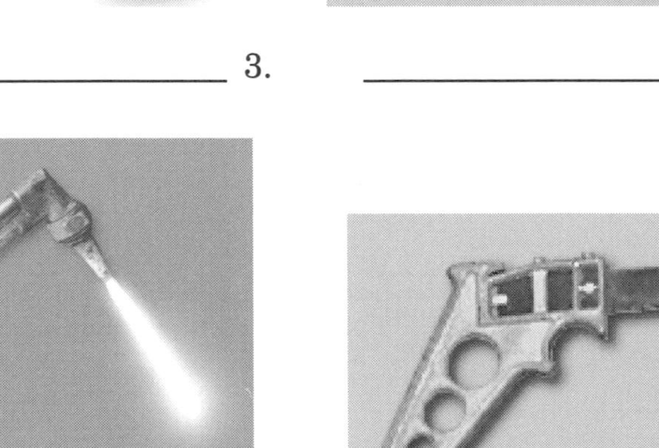

______________________ 5.

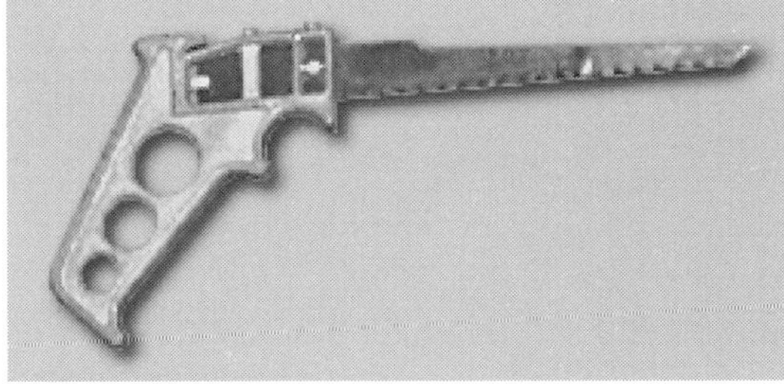

______________________ 6.

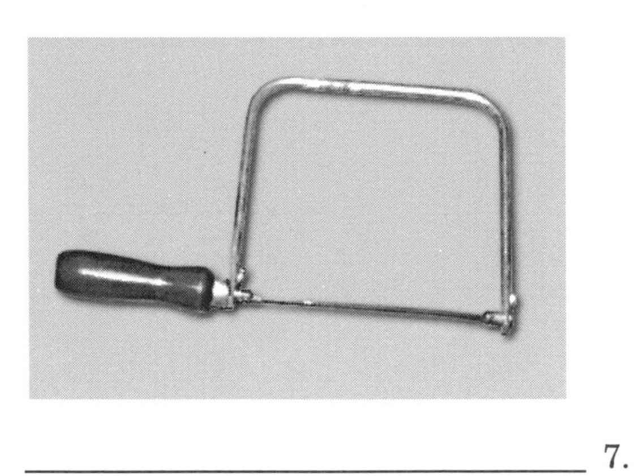

______________________ 7.

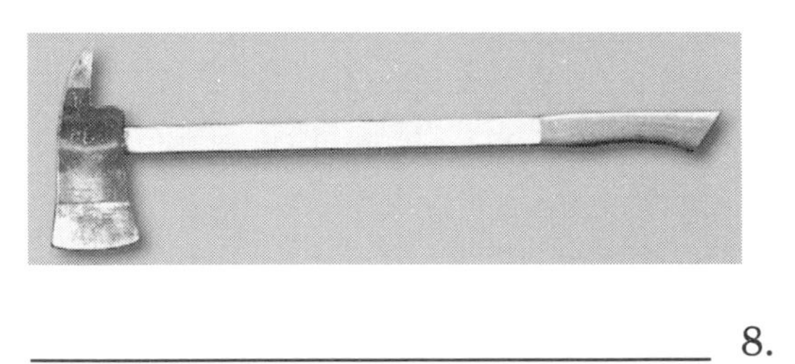

______________________ 8.

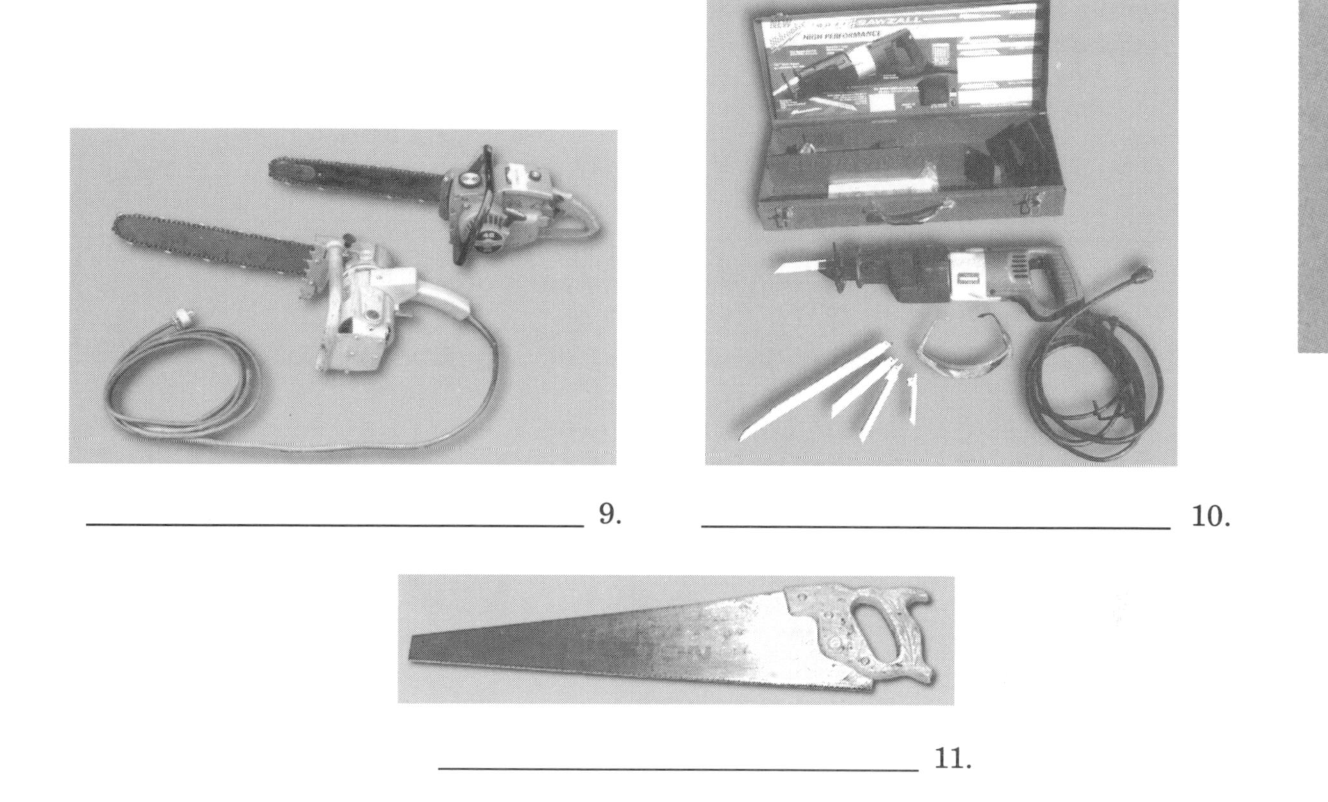

_______________ 9.

_______________ 10.

_______________ 11.

E. **Identify the following pushing/pulling tools used for forcible entry. Write the correct name below each.**

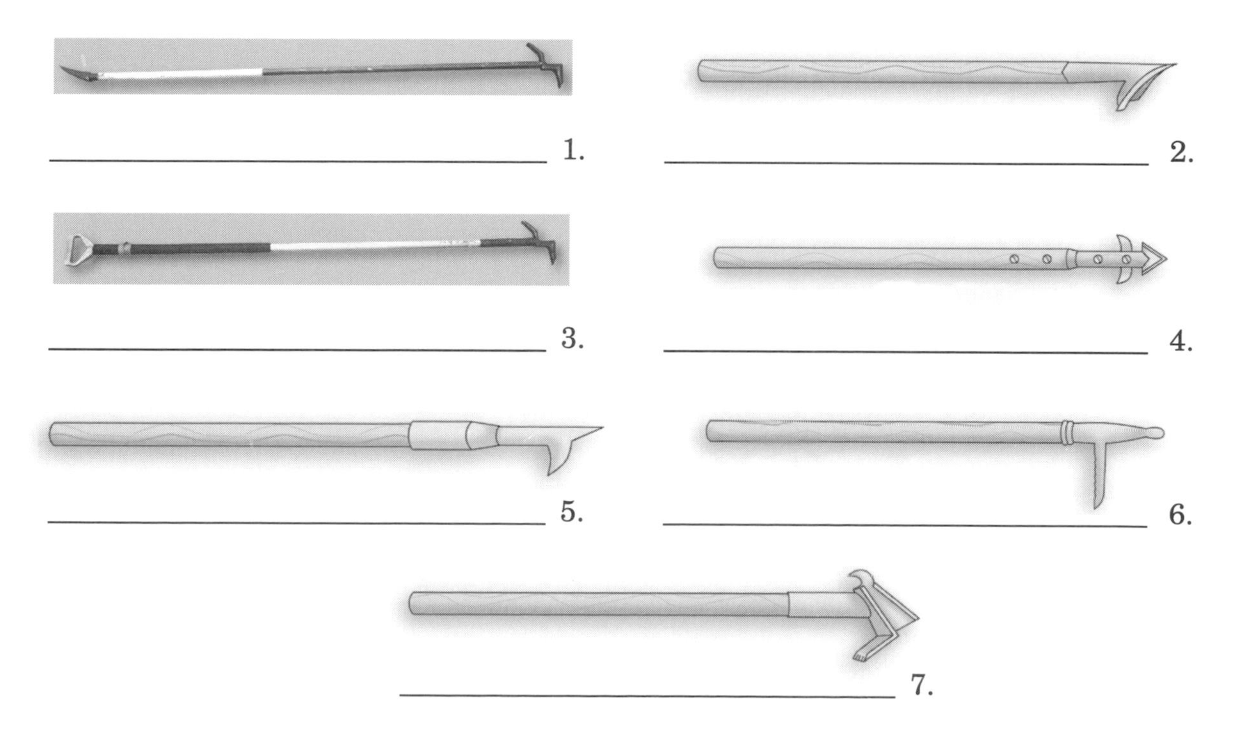

_______________ 1.

_______________ 2.

_______________ 3.

_______________ 4.

_______________ 5.

_______________ 6.

_______________ 7.

Photo for Question 10 courtesy of Keith Flood.

F. **Identify the following prying tools used for forcible entry. Write the correct name below each.**

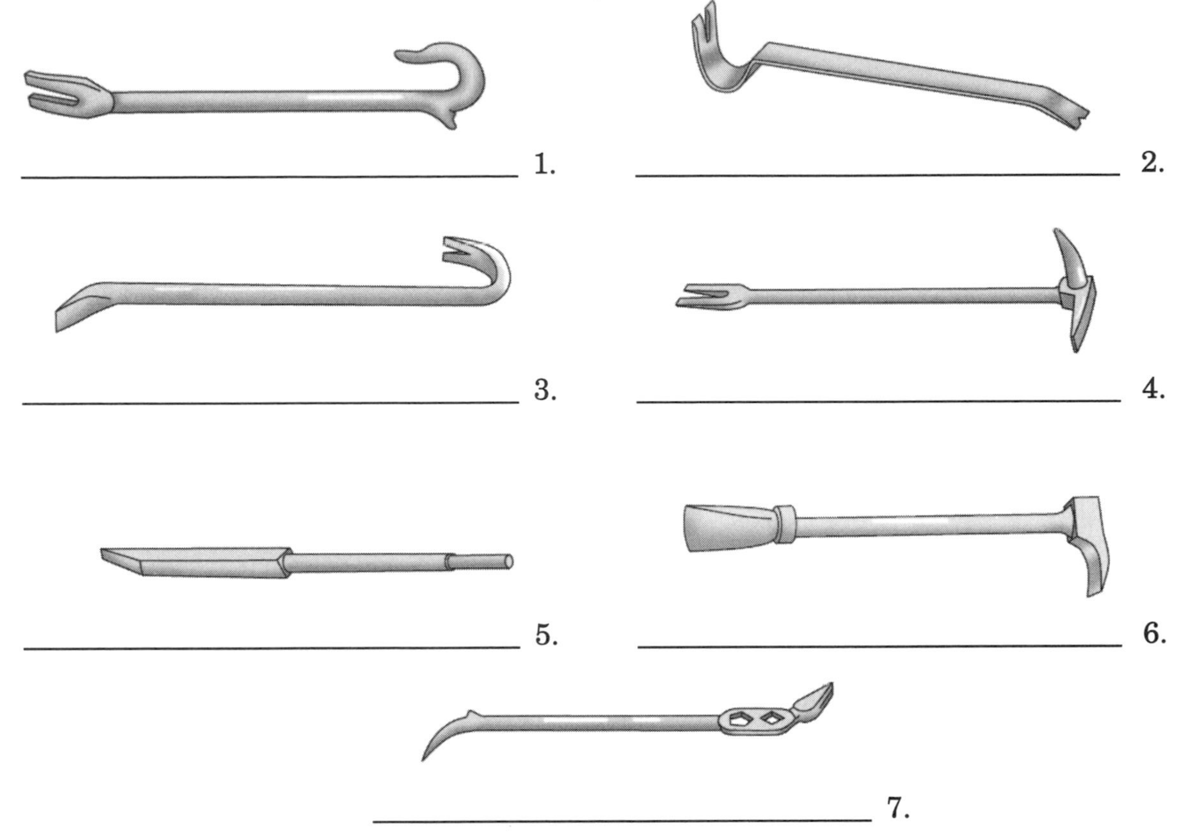

_______________ 1.

_______________ 2.

_______________ 3.

_______________ 4.

_______________ 5.

_______________ 6.

_______________ 7.

G. **Identify the following striking tools used for forcible entry. Write the correct name below each.**

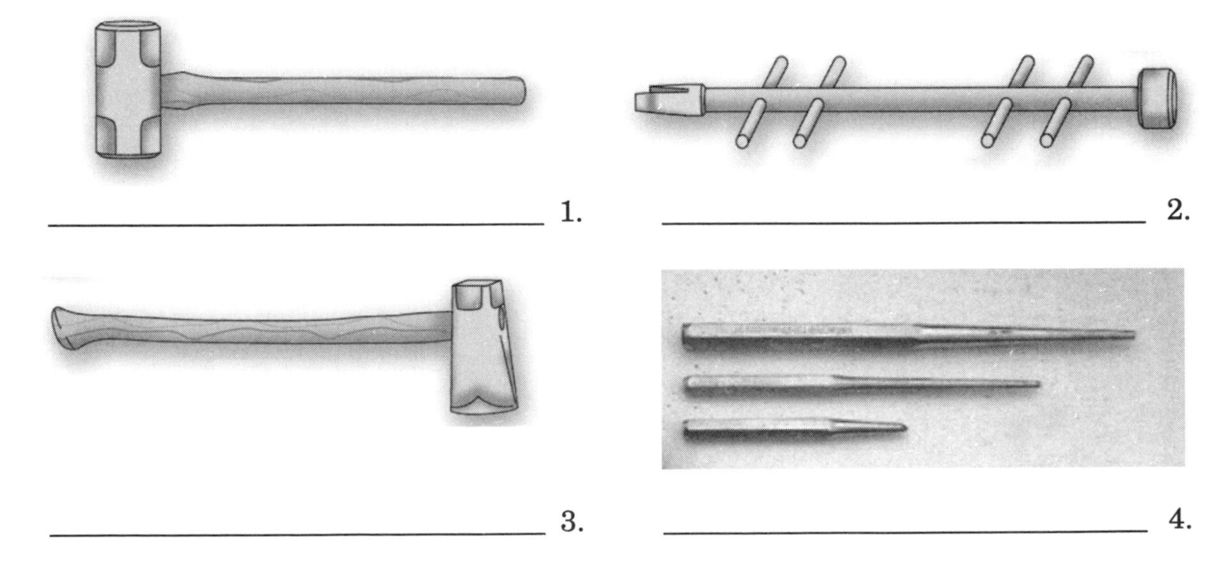

_______________ 1.

_______________ 2.

_______________ 3.

_______________ 4.

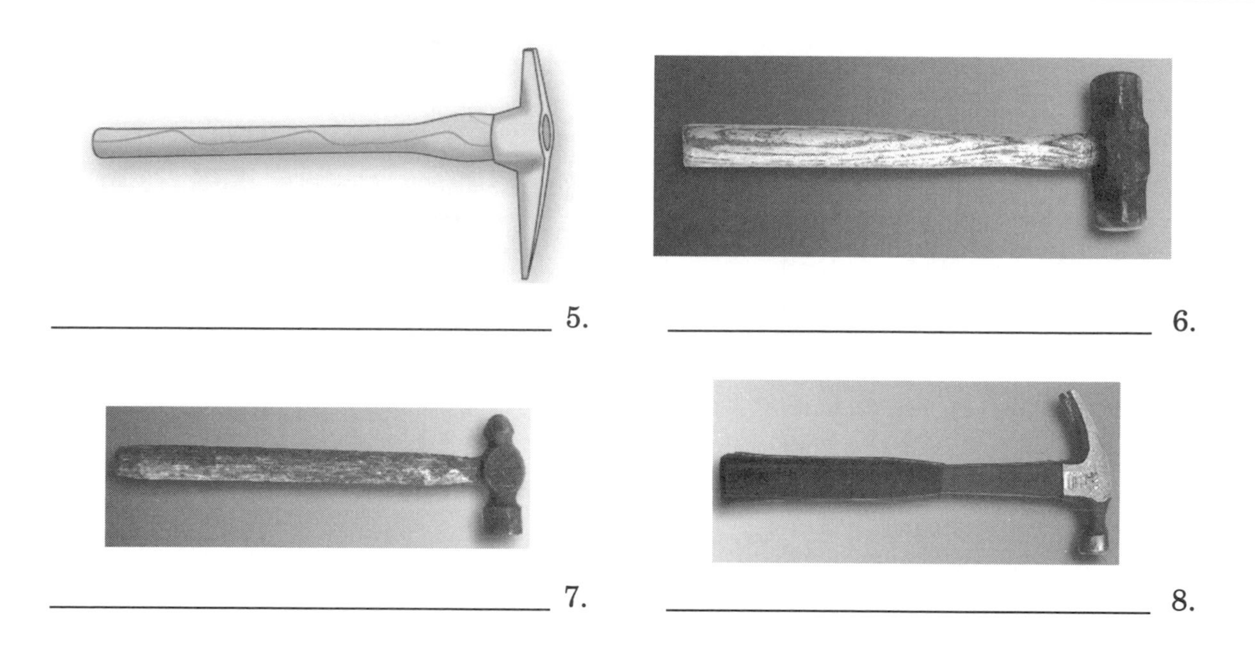

_______________ 5.

_______________ 6.

_______________ 7.

_______________ 8.

H. **Identify the following types of doors and jambs. Write the correct name below each.**

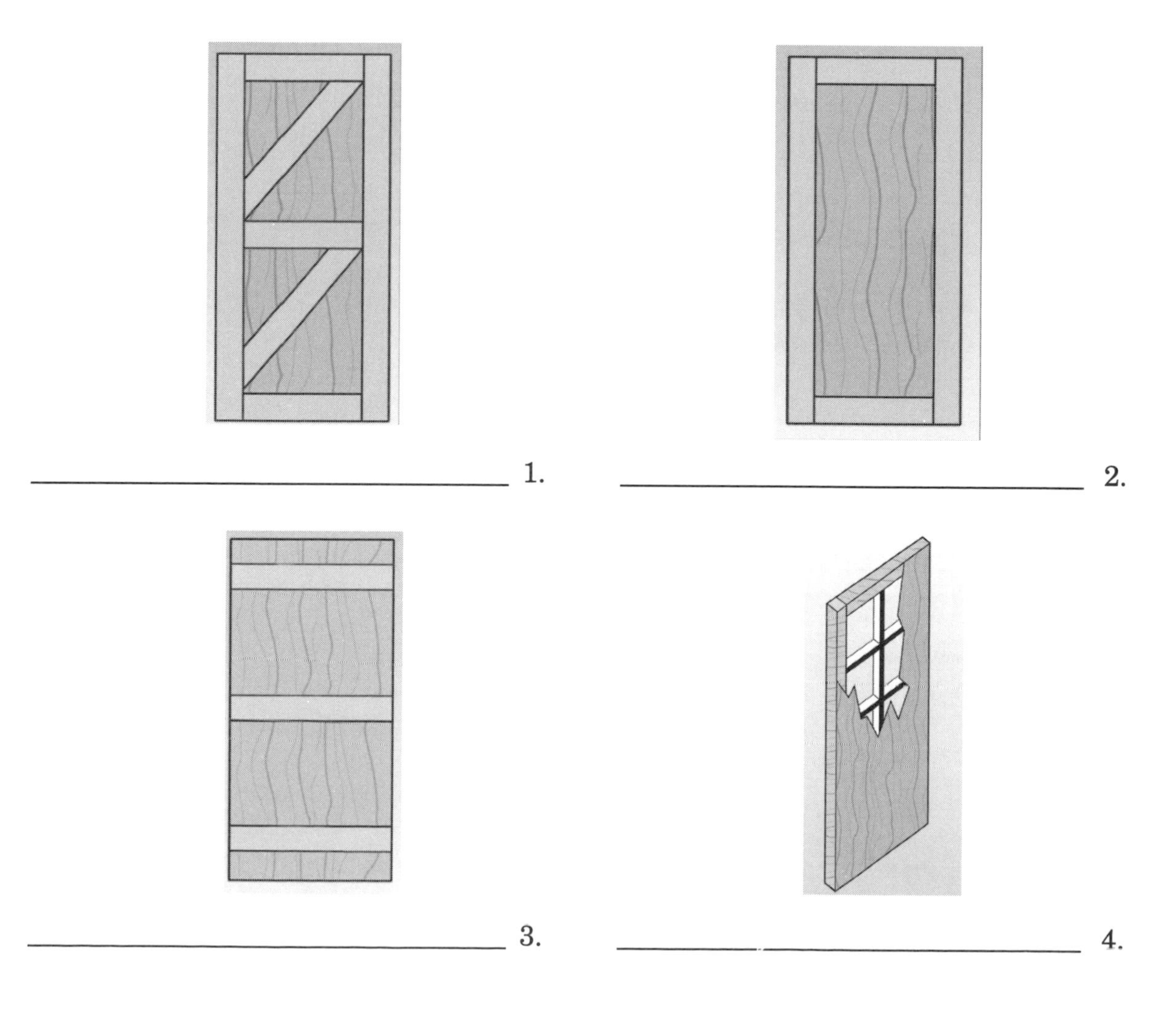

_______________ 1.

_______________ 2.

_______________ 3.

_______________ 4.

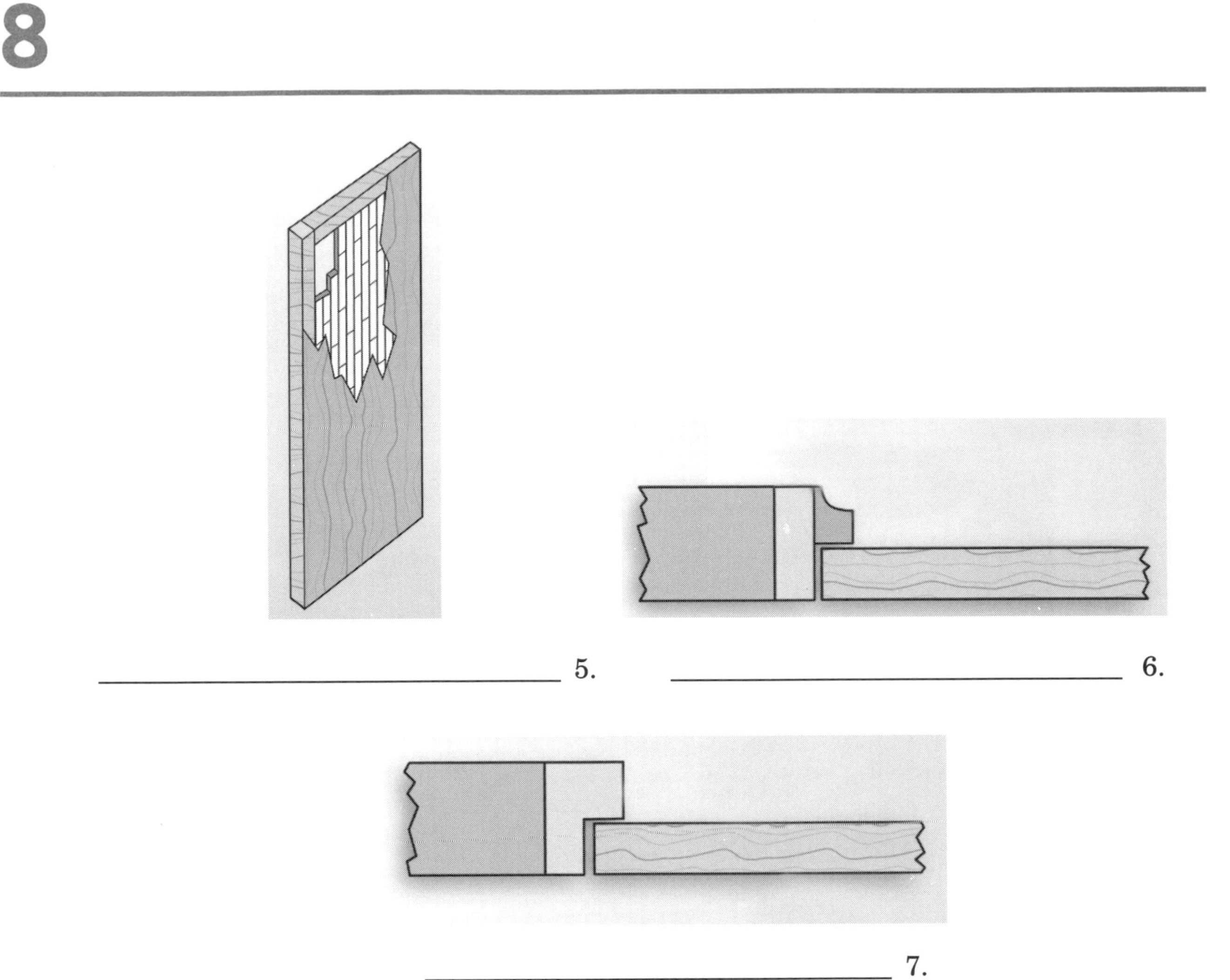

_______________ 5.

_______________ 6.

_______________ 7.

I. Identify the following types of sliding, revolving, and overhead doors. Write the correct name below each.

_______________ 1.

_______________ 2.

_______________ 3.

_______________ 4.

J. **Identify the following locks and locking devices. Write the correct name below each.**

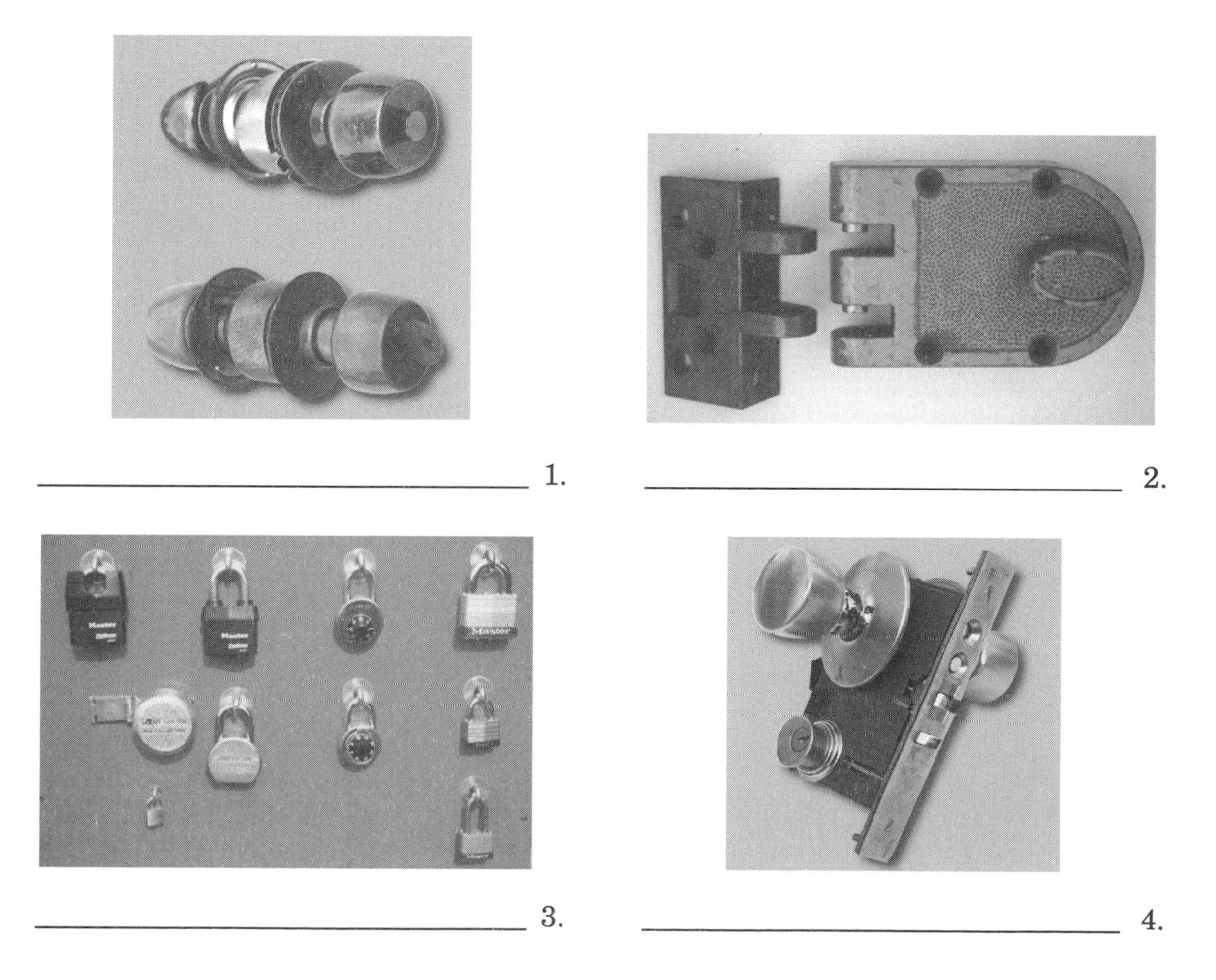

_______________ 1.

_______________ 2.

_______________ 3.

_______________ 4.

K. Identify the following tools for through-the-lock forcible entry. Write the correct name below each.

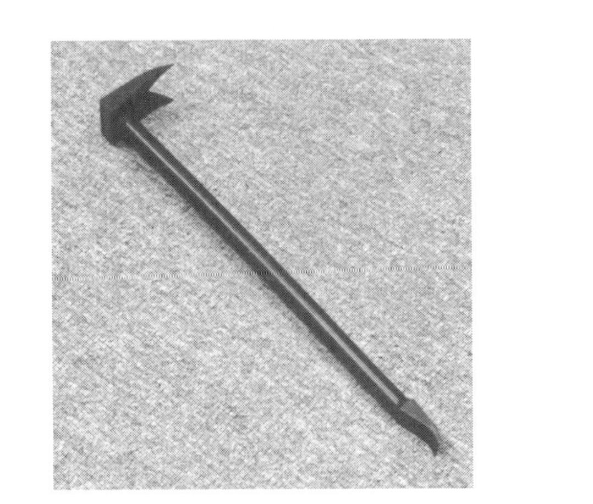

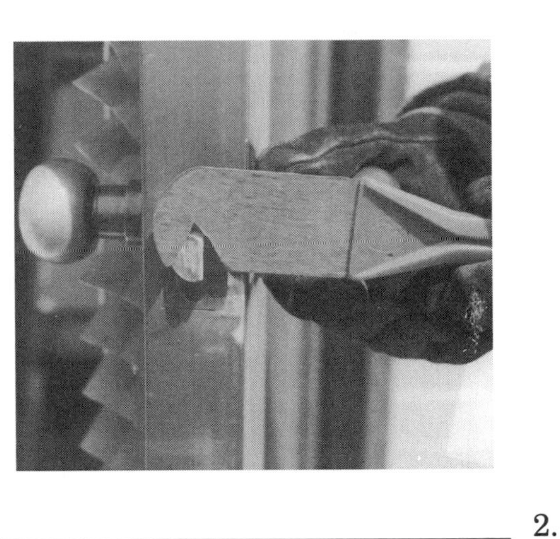

________________________ 1. ________________________ 2.

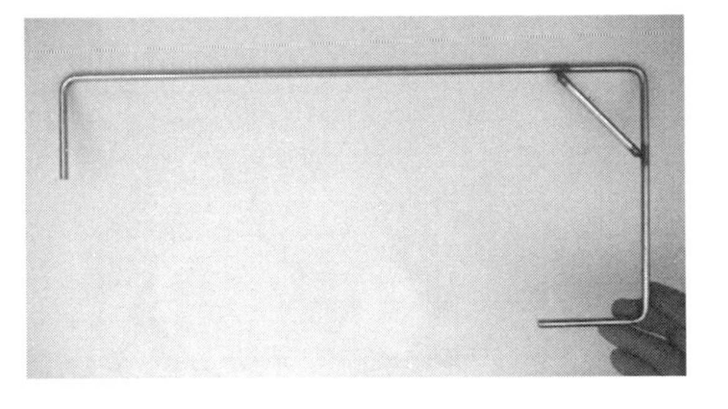

________________________ 3.

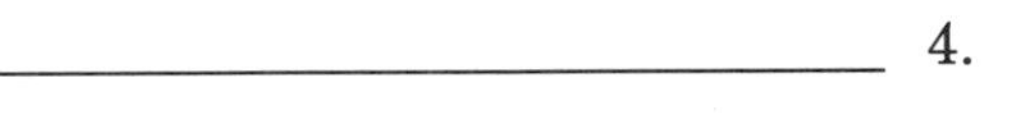

________________________ 4.

L. Identify the following tools for breaking padlocks. Write the correct name below each.

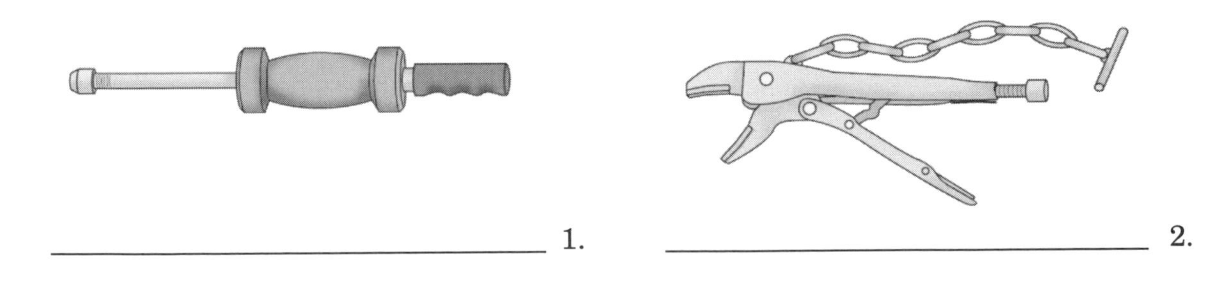

________________________ 1. ________________________ 2.

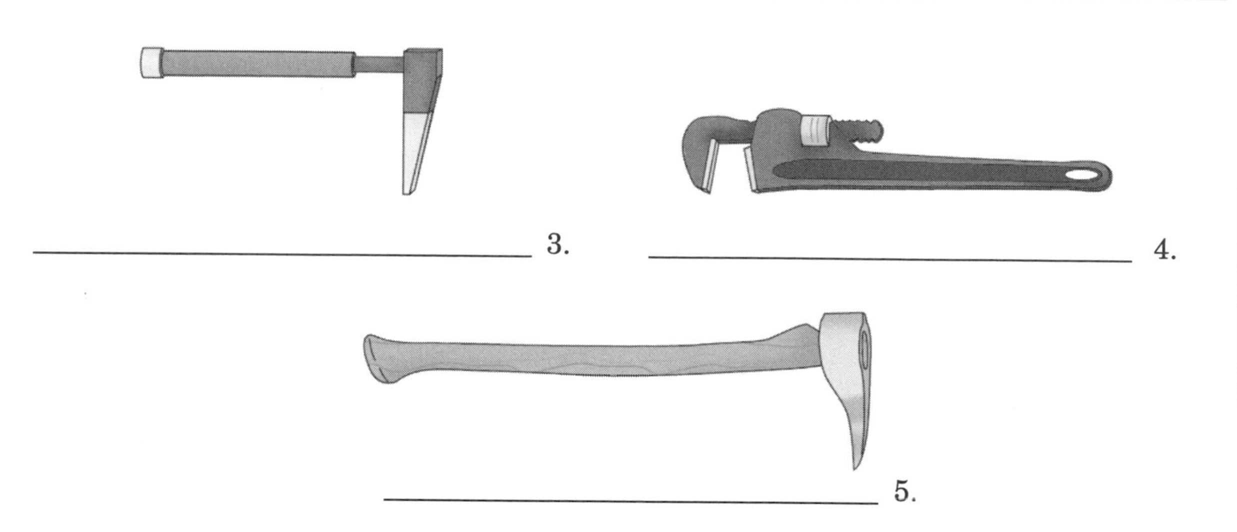

______________________ 3. ______________________ 4.

______________________ 5.

M. **Identify the following types of windows. Write the correct name below each.**

______________________ 1. ______________________ 2.

__________________________ 3.

__________________________ 4.

__________________________ 5.

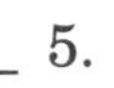

Chapter 9
Ground Ladders

Chapter 9 Ground Ladders

FIREFIGHTER I

Matching

A. Match ladder parts to their functions. Write the appropriate letters on the blanks. Functions are continued and ladder parts are repeated on next page.

____ 1. Main structural member of a ladder supporting the rungs or blocks

____ 2. Lowest or widest section of an extension ladder; this section always maintains contact with the ground or other supporting surface

____ 3. Bottom end of a ladder; the end that is placed on the ground or other supporting surface when the ladder is raised

____ 4. Metal safety plates or spikes attached to the butt of a ground ladder's beams to prevent slippage

____ 5. Upper section(s) of extension or some combination ladders

____ 6. Wood or metal strips, sometimes in the form of slots or channels, on an extension ladder that guide the fly section while being raised

____ 7. Rope or cable used for hoisting and lowering the fly sections of an extension ladder

____ 8. Curved metal devices installed on the tip end of roof ladders to secure the ladder to the highest point on the roof of a building

____ 9. Devices attached to the inside of the beams on fly sections used to hold the fly section in place after it has been extended

a. Butt spurs
b. Hooks
c. Beam
d. Halyard
e. Protection plates
f. Guides
g. Tie rods
h. Bed section (base section)
i. Stops
j. Pawls (dogs)
k. Truss block
l. Fly
m. Rails
n. Footpads
o. Butt (heel)

____ 10. Strips of metal attached to ladders at chafing points

____ 11. Two lengthwise members of a trussed ladder beam that are separated by truss or separation blocks

____ 12. Wood or metal pieces that prevent the fly section from being extended too far

____ 13. Metal rods running from one beam to the other

____ 14. Separation pieces between the rails of a trussed ladder; sometimes used to support rungs

a. Butt spurs
b. Hooks
c. Beam
d. Halyard
e. Protection plates
f. Guides
g. Tie rods
h. Bed section (base section)
i. Stops
j. Pawls (dogs)
k. Truss block
l. Fly
m. Rails
n. Footpads
o. Butt (heel)

B. Match ladder types to their descriptions. Write the appropriate letters on the blanks.

____ 1. Nonadjustable in length and consists of only one section

____ 2. Single ladders equipped at the tip with folding hooks

____ 3. Single ladders that have hinged rungs allowing them to be folded so that one beam rests against the other

____ 4. Adjustable in length and consists of a base or bed section and one or more fly sections

____ 5. Designed for use as a self-supported stepladder (A-frame) and as a single or extension ladder

____ 6. Single-beam ladder with rungs projecting from both sides

a. Extension
b. Roof
c. Step
d. Pompier (scaling)
e. Single (wall)
f. Combination
g. Folding

True/False

C. Write *True* or *False* before each of the following statements. Correct those statements that are false.

_______ 1. According to NFPA standards, folding ladders must have footpads attached to the butt to prevent slipping.

_______ 2. Before using a ladder *without* a manufacturer's certification label, firefighters must thoroughly test the ladder for safety.

_______ 3. Wood ladders with signs of minimal wood deterioration may still be used, but the deterioration should be noted for the next inspection.

_______ 4. Ladders that cannot be safely repaired should be used only for station work.

_______ 5. Firefighters must wear gloves while working on ladders.

_______ 6. Metal fire service ladders are specially insulated to prevent firefighters from being electrocuted if a ladder comes into contact with electrical lines.

_______ 7. The reach of a ladder is equivalent to the designated length indicated on the ladder.

____

____

_______ 8. The NFPA standard for ground ladders designates the location and methods of mounting ground ladders on fire apparatus.

____

____

_______ 9. Firefighters should use the strength of their legs, not their arms, to lift and lower ladders.

____

____

_______ 10. Firefighters should place ladders at two or more points on a fire building.

____

____

_______ 11. If the butt of a ladder is too close to a building, the load-carrying capacity of the ladder is reduced, and it has more of a tendency to slip.

____

____

_______ 12. A firefighter should place an extension ladder flat on the ground prior to raising it.

____

____

_______ 13. Fiberglass ladders are safe to use near live electrical wires.

____

____

_______ 14. The recommended safe distance between ladders and energized electrical lines applies to ladder placement, not ladder raising.

____

____

_______ 15. Space permitting, a firefighter can raise a ladder either parallel with or perpendicular to a building.

____

____

_______ 16. When a roof ladder is carried to the scene butt first, firefighters should turn the ladder around before placing it.

____

____

_______ 17. A firefighter can secure a ladder for climbing by standing on the outside of the ladder and chocking the butt end with his/her feet.

____

____

_______ 18. If a ladder is properly placed against a building, the firefighter's body will lean slightly toward the building when standing on the ladder.

____

____

_______ 19. Firefighters should climb ladders while keeping their eyes focused forward and occasionally glancing up.

____

____

_______ 20. Firefighters should keep their arms bent and their bodies close to the ladder while climbing.

____

____

_______ 21. Firefighters should focus on developing speed first during ladder-climbing practice sessions.

____

____

_______ 22. Firefighters should not exceed a ladder's rated capacity by allowing more than one person on each section at the same time.

____

____

_______ 23. Firefighters should lower conscious victims feet first from a building onto a ladder.

____

____

Multiple Choice

D. Write the letter of the best answer on the blank before each statement.

____ 1. What NFPA standard contains the requirements for ground ladder design and manufacturer testing?

a. 1901
b. 1931
c. 1932
d. 1941

____ 2. Which of the following statements is true about single ladders?

a. They provide quick access to windows.
b. They have moderate strength and moderate weight.
c. They are available in various lengths from 10 to 30 feet *(3 m to 9 m)*.
d. They should not be made of truss construction.

____ 3. What is the length range of roof ladders?

a. 8 to 16 feet *(2.5 m to 5 m)*
b. 10 to 18 feet *(3 m to 5.5 m)*
c. 12 to 20 feet *(4 m to 6 m)*
d. 12 to 24 feet *(4 m to 8 m)*

____ 4. What is the most common length of folding ladders?

a. 10 feet *(3 m)*
b. 12 feet *(4 m)*
c. 14 feet *(4.3 m)*
d. 16 feet *(5 m)*

____ 5. On what type of ladder is a stop found?

a. Single
b. Folding
c. Combination
d. Extension

____ 6. For what task is the pompier ladder used?

a. Rescuing victims of trench collapse
b. Climbing from floor to floor via exterior windows on multistory buildings
c. Obtaining quick access to windows and roofs on one- and two-story buildings
d. Accessing attics and scuttle holes in small rooms and closets

____ 7. What is the general length range of extension ground ladders?

a. 12 to 39 feet *(4 m to 11.5 m)*
b. 18 to 45 feet *(5.5 m to 14 m)*
c. 20 to 60 feet *(6 m to 18 m)*
d. 22 to 42 feet *(6.7 m to 12.8 m)*

____ 8. What NFPA-required mechanisms hold a combination ladder in the open position?

a. Rigid pawls
b. Positive locking devices
c. Protection plates
d. Steel rod ties

____ 9. In which of the following ways should a combination ladder *not* be used?

a. As a single ladder
b. As a pompier ladder
c. As an extension ladder
d. As an A-frame ladder

____ 10. What fire service term means *to keep ladders in a state of usefulness or readiness?*

a. Repair
b. Restoration
c. Maintenance
d. Upkeep

____ 11. Which of the following guidelines applies to the care of all types of ground ladders?

a. Store ground ladders on exterior fire station hooks.
b. Paint the top and bottom two rungs of ladders for identification and visibility purposes.
c. Keep ladders away from apparatus exhaust systems.
d. Clean ladders with turpentine or gasoline after each use.

____ 12. What should be used to remove tar, oil, and greasy residues from ladders?

a. Distilled water
b. Safety solvents
c. Gasoline
d. Soapy water

____ 13. What are the most effective tools for cleaning dirt and debris from ladders?

a. A sponge and soapy water
b. A soft-bristled brush and turpentine
c. A soft-bristled brush and running water
d. A high-pressure nozzle and running water

____ 14. What does bubbled varnish on a wood ladder usually indicate?

a. Water damage
b. Inferior workmanship
c. Normal wear
d. Heat exposure

____ 15. What circumstance would *not* be a reason to place a metal ladder out of service until tested?

a. A ladder coming into contact with a direct flame
b. The heat sensor label changing color
c. A weld showing signs of minimal deterioration
d. A ladder becoming hot to the touch from heat exposure

____ 16. What do dark streaks on wood ladders indicate?
a. Deterioration of the wood
b. Especially strong wood
c. Deterioration of metal parts
d. Natural aging and seasoning of the wood

____ 17. What does the darkening of varnish on wood ladders indicate?
a. Deterioration from overuse
b. Natural aging of the varnish
c. Heat exposure
d. Time to replace the varnish

____ 18. What condition indicates that an extension ladder is not safe to use?
a. The hook and finger move in and out freely.
b. The halyard cable has slack when the ladder is in the bedded position.
c. The pole ladder staypole toggles operate unchecked.
d. The fly sections of the ladder move smoothly.

____ 19. Which of the following ladders is *not* recommended by NFPA as standard equipment on fire department pumpers?
a. One extension ladder at least 24 feet *(8 m)* long
b. One roof ladder at least 14 feet *(4.3 m)* long
c. One combination ladder at least 14 feet *(4.3 m)* long
d. One folding ladder at least 10 feet *(3 m)* long

____ 20. What ladder length would be best to reach a typical first-story roof?
a. 10 to 15 feet *(3 m to 4.6 m)*
b. 14 to 18 feet *(4 m to 5.5 m)*
c. 16 to 20 feet *(4.9 m to 6 m)*
d. 20 to 28 feet *(6 m to 8.5 m)*

____ 21. What ladder length would be best to reach a third-story window or roof?
a. 16 to 20 feet *(4.9 m to 6 m)*
b. 20 to 28 feet *(6 m to 8.5 m)*
c. 28 to 40 feet *(8.5 m to 12.2 m)*
d. 40 to 50 feet *(12.2 m to 15.2 m)*

____ 22. How far beyond a roof edge should ladders extend to provide both a footing and a handhold for persons stepping on or off the ladder?
a. 2 rungs
b. 3 rungs
c. 4 rungs
d. 5 rungs

____ 23. Where should the tip of a ladder be placed for gaining access from the side of a window or for ventilation?
a. Even with the top of the window
b. Between the windowsill and the top of the window
c. Just below the windowsill
d. Above the windowsill by two rungs

____ 24. Where should the tip of the ladder be placed for rescues from a window opening?

a. Even with the top of the window
b. Between the windowsill and the top of the window
c. Just below the windowsill
d. Above the windowsill by two rungs

____ 25. What is the maximum extended length of extension ladders?

a. Approximately 5 inches *(125 mm)* more than the specified length
b. As much as 6 inches *(150 mm)* less than the designated length
c. Up to 8 inches *(200 mm)* more than the projected length
d. About 10 inches *(250 mm)* less than the determined length

____ 26. What is the maximum reach or working height for a 16-foot *(4.9 m)* ladder placed at the proper climbing angle?

a. 13 feet *(4 m)*
b. 14 feet *(4.3 m)*
c. 15 feet *(4.6 m)*
d. 16 feet *(4.9 m)*

____ 27. Who should give the lift command when two or more firefighters are lifting a ladder?

a. Firefighter at the rear
b. Firefighter at the front
c. Firefighter in the middle
d. Firefighter with the most experience

____ 28. What carry should a firefighter use to transport a single ladder alone?

a. Flat-shoulder
b. Low-shoulder
c. Arm's length on-edge
d. Over-the-shoulder

____ 29. What carry is typically used for extension ladders up to 35 feet *(10.7 m)?*

a. Three-firefighter flat-shoulder
b. Two-firefighter low-shoulder
c. Three-firefighter arm's length on-edge
d. Four-firefighter over-the-shoulder

____ 30. How should firefighters position the forward end of a ladder during a carry?

a. Parallel to the ground
b. Slightly raised
c. Slightly lowered
d. Either slightly raised or parallel to the ground

____ 31. For what ladder is the two-firefighter low-shoulder carry most often used?

a. Single
b. Roof
c. Pompier
d. Extension

____ 32. What ladder end should firefighters face after completing a two-firefighter low-shoulder carry from flat on the ground?
   a. Tip of the ladder
   b. Butt of the ladder
   c. Either end
   d. The end determined during size-up

____ 33. How should firefighters carry a roof ladder that needs to be carried up a ground ladder and placed with hooks deployed on a sloped roof?
   a. Low-shoulder with the tip (hooks) forward
   b. Low-shoulder with the tip (hooks) behind
   c. Flat-shoulder with the tip (hooks) forward
   d. Flat-shoulder with the tip (hooks) behind

____ 34. How should a firefighter carry a roof ladder when there is no second firefighter to open the hooks, time is critical, and there is no crowd of people through which the ladder must be carried?
   a. With the hooks closed and turned inward in relation to the firefighter carrying the ladder
   b. With the hooks closed and turned outward in relation to the firefighter carrying the ladder
   c. With the hooks opened and turned inward in relation to the firefighter carrying the ladder
   d. With the hooks opened and turned outward in relation to the firefighter carrying the ladder

____ 35. Who typically decides the *exact* location where the butt of a ground ladder will be placed?
   a. Commanding officer
   b. Incident commander
   c. Firefighter nearest the tip of the ladder
   d. Firefighter nearest the butt of the ladder

____ 36. How should a ladder be placed so that firefighters can climb in or out narrow windows?
   a. Alongside the window to the downwind side
   b. Just beneath the window
   c. With the tip about even with the top of the window
   d. With two or three rungs above the windowsill

____ 37. When should firefighters raise a ladder directly in front of a window with the tip on the wall above the window opening?
   a. When used for entry
   b. When used as a smoke fan support
   c. When used for rescue operations
   d. When used for ventilation operations

____ 38. What is the desired angle of inclination when a ladder has been raised and lowered into place?

a. 60 degrees
b. 65 degrees
c. 75 degrees
d. 80 degrees

____ 39. How can firefighters determine the proper number of feet *(meters)* between the heel of the ladder and the building?

a. Divide the used length of the ladder by 4
b. Subtract 4 feet *(1.2 m)* for every 6-foot *(1.8 m)* ladder section
c. Multiply the length of the ladder by 0.4
d. Add 1 foot *(0.3 m)* for every 10 ladder rungs

____ 40. What distance should be maintained between a ladder and energized electrical lines?

a. 4 feet *(1.2 m)*
b. 6 feet *(1.8 m)*
c. 10 feet *(3 m)*
d. 13 feet *(4 m)*

____ 41. How should the fly section on an extension ladder be positioned?

a. With the fly in (toward the building)
b. With the fly out (away from the building)
c. Consistent with the experience level of the firefighters
d. Consistent with the manufacturer's specifications

____ 42. What should be done with the excess halyard once an extension ladder is resting against a building and before the ladder is climbed?

a. Tied to the ladder with a clove hitch
b. Clamped to the rail of the ladder
c. Allowed to hang between the ladder and the building
d. Wound snugly around the lowest rung

____ 43. For which of the following ladders can a firefighter place the butt so that the ladder can be climbed without heeling it against the building?

a. 15-foot *(4.6 m)* single
b. 12-foot *(4 m)* extension
c. 14-foot *(4.3 m)* roof
d. 10-foot *(3 m)* pompier

____ 44. Who decides whether a ladder will be raised parallel with or perpendicular to a building?

a. An officer at the scene
b. The firefighter at the butt of the ladder
c. The firefighter at the tip of the ladder
d. The firefighter with the most ladder experience

____ 45. How many firefighters should typically work together to raise a ladder of 35 feet *(10.7 m)* or longer?

a. Two
b. Three
c. Four
d. Five

____ 46. What raise should firefighters use for a large, heavy extension ladder?
a. Flat
b. Beam
c. Butt-first
d. Pivot

____ 47. What should firefighters do if a ladder is raised with the fly in the incorrect position for deployment?
a. Lower the ladder, lay it flat on the ground, and rotate it.
b. Pull the ladder away from the building, lift it off the ground, and rotate it.
c. Leave the ladder in the position it was raised, and use it normally.
d. Place a foot against a beam, tilt the ladder onto the beam, and pivot it.

____ 48. Which vertically positioned ladder can be safely shifted by one firefighter?
a. 12-foot *(4 m)* extension
b. 15-foot *(4.6 m)* pompier
c. 20-foot *(6 m)* single
d. 24-foot *(8 m)* roof

____ 49. What ladder part should a firefighter grasp when heeling a ladder?
a. Beam
b. Middle rung
c. Butt
d. Bottom rung

____ 50. How can a firefighter cause the least possible bounce and sway when ascending a ladder?
a. Keep legs straight.
b. Keep knees bent.
c. Hold body close to the ladder.
d. Climb quickly.

____ 51. How should firefighters maintain the correct distance between themselves and ladders while climbing?
a. Reach upward during the climb.
b. Lean against the ladder.
c. Hold arms straight.
d. Drape one arm over rungs at the elbow joint.

____ 52. What should firefighters *not* use to secure themselves to ladders?
a. Class I safety harness
b. Leg lock
c. Ladder belt
d. Class III safety harness

____ 53. Which of the following methods would *not* be appropriate for assisting an unconscious victim down a ladder?
a. Resting the victim's body on the rescuer's supporting knee with the victim's feet placed outside the rails
b. Lowering the victim feet first with one rescuer holding the victim's armpits and the other rescuer holding the victim's knees
c. Supporting the victim at the crotch with one arm and at the chest with the other arm
d. Cradling the victim in front of the rescuer with the victim's legs over the rescuer's shoulders and the victim's arms draped over the rescuer's arms

____ 54. For what victim would rescuers use two ground ladders placed side by side?
- a. An elderly, thin, unconscious woman
- b. An extremely heavy, conscious man
- c. A young, unconscious girl
- d. A chubby, 7-year-old, conscious boy

Identify

E. Identify the following ground ladders. Write the correct name below each.

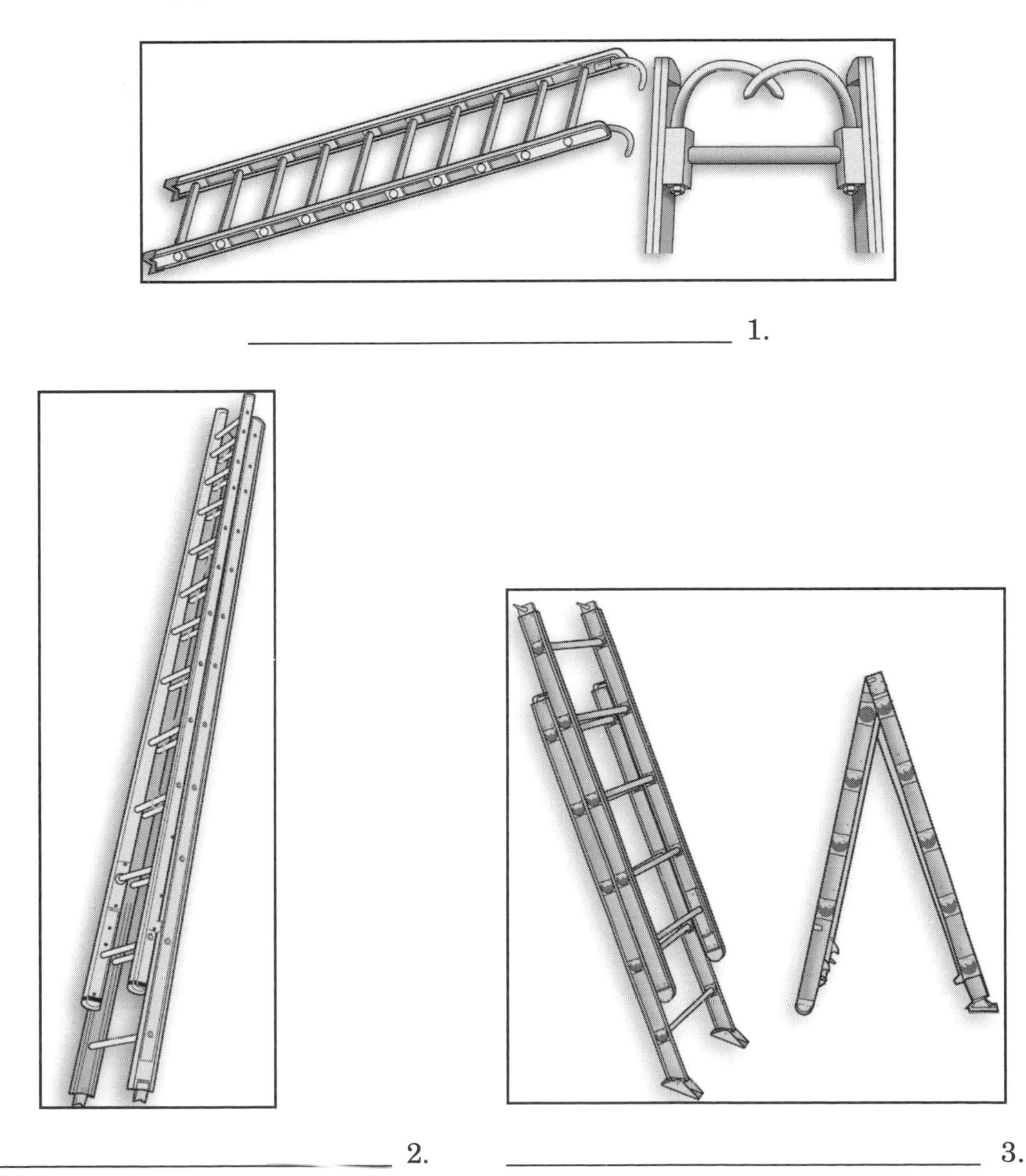

________________________ 1.

________________________ 2.

________________________ 3.

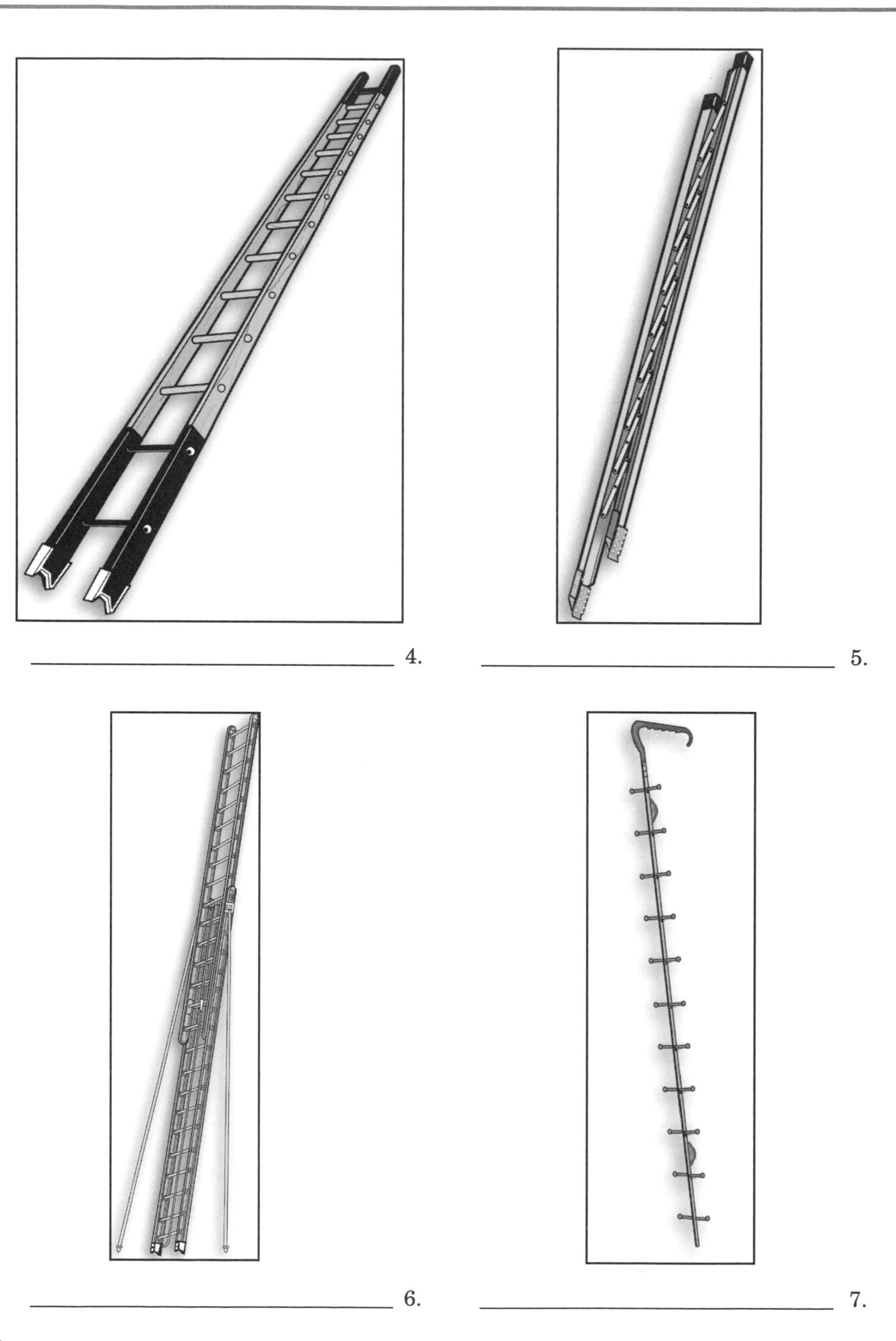

_________________________ 4.

_________________________ 5.

_________________________ 6.

_________________________ 7.

Label

F. Label the following ladder parts. Write the correct names on the blanks next to the corresponding letters.

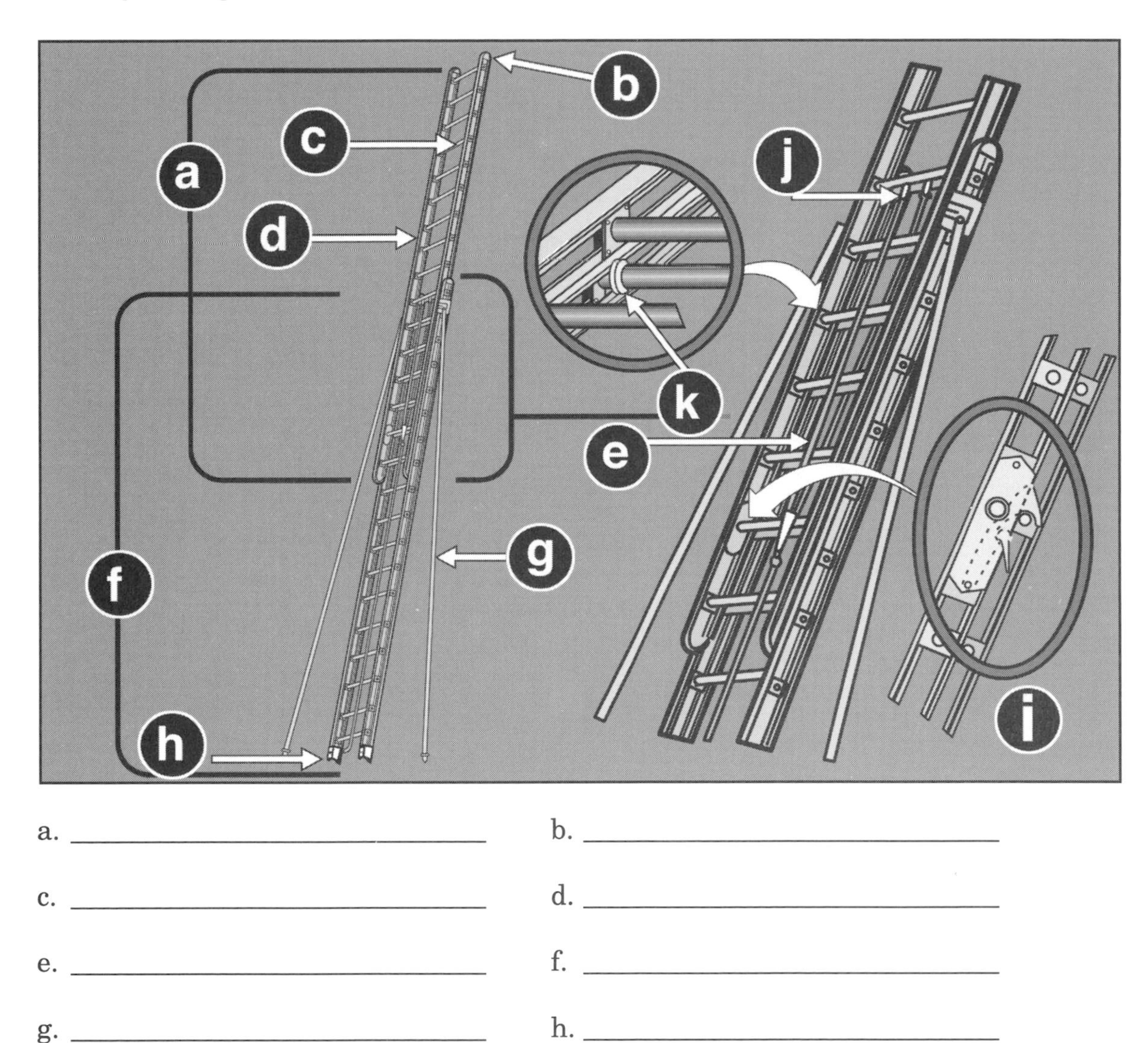

a. ______________________ b. ______________________

c. ______________________ d. ______________________

e. ______________________ f. ______________________

g. ______________________ h. ______________________

i. ______________________ j. ______________________

k. ______________________

Chapter 10
Ventilation

Chapter 10 Ventilation

FIREFIGHTER I

Matching

A. Match to their definitions terms associated with ventilation. Write the appropriate letters on the blanks.

____ 1. The systematic removal and replacement with cooler air of heated air, smoke, and gases from a structure

____ 2. Exposed part of a roof that protects a structure from weather

____ 3. The drawing of air currents from throughout a building in the direction of an opening

____ 4. Phenomenon in which accumulated heat, smoke, and fire gases bank down and spread laterally to involve other areas of a structure

____ 5. Phenomenon that may occur during the transition between the fire growth and the fully developed fire stages

____ 6. Side of a building the wind is striking

____ 7. Air recirculating around the sides of a fan and in and out of nearby openings

____ 8. Location where positive-pressure ventilation is done

a. Chimney effect
b. Ventilation
c. Leeward
d. Windward
e. Flashover
f. Mushrooming
g. Roof covering
h. Point of entry
i. Churning action

B. Match ventilation types to their definitions. Write the appropriate letters on the blanks.

____ 1. Opening a 4-foot *(1.2 m)* wide roof area the width of the building to channel out fire and heat

____ 2. Venting heat, smoke, and gases through wall openings such as windows and doors

____ 3. Mechanically or hydraulically venting a structure

____ 4. Using fans to develop artificial circulation and to pull smoke out of a structure

____ 5. Mechanically blowing fresh air into a structure in sufficient volume to create a higher pressure within the structure and thereby forcing the contaminated atmosphere out the exit opening

____ 6. Directing a water fog stream out a window to increase air and smoke movement

a. Horizontal
b. Hydraulic
c. Trench
d. Negative-pressure
e. Forced
f. Hydration
g. Positive-pressure

True/False

C. Write *True* or *False* before each of the following statements. Correct those statements that are false.

________ 1. Ventilation reduces the chances of flashover and backdraft.

____

____

________ 2. Insulation installed over fire-rated roof coverings increases the fire rating of the roof construction.

____

____

_______ 3. Use a water or fog spray to dissipate, absorb, or expel fire gases and smoke.

____

____

_______ 4. Use stairwells for evacuation and ventilation simultaneously to ensure life safety.

____

____

_______ 5. Do not work with the wind to your *side* while cutting a roof opening.

____

____

_______ 6. Remain on a roof to monitor ventilation for at least 5 minutes after completing ventilation work.

____

____

_______ 7. Personnel performing roof ventilation work must wear goggles.

____

____

_______ 8. Do not angle a ventilation cut toward the body.

____

____

_______ 9. Most existing roof openings are unlocked and easy to access.

____

____

_______ 10. Firefighters usually accomplish ventilation tasks quicker by cutting a hole in a roof than by trying to force a locked or secured existing roof opening.

____

____

_______ 11. The structural part of a flat roof is similar to the construction of a floor that consists of wooden, concrete, or metal joists covered with sheathing.

_______ 12. Slate and tile pitched roofs must be opened with a rotary saw.

_______ 13. Use extreme caution when climbing onto a roof that contains a significant amount of fire in the truss area.

_______ 14. Firefighters can cut a hole of considerable size through a trussless roof without causing the roof structure to collapse.

_______ 15. Lightweight concrete slabs can be opened with a sledgehammer.

_______ 16. Operating a fire stream through a ventilation hole during offensive operations aids the ventilation process.

_______ 17. Because horizontal ventilation does not normally release heat and smoke directly above the fire, some routing is necessary.

_______ 18. Horizontal ventilation eliminates the danger that rising heat and gases will ignite higher portions of a fire building.

_____ 19. To perform negative-pressure ventilation, place a fan to exhaust in the same direction as the natural wind.

____

____

_____ 20. Do not use forced-air fans in flammable atmospheres.

____

____

_____ 21. As long as the pressure is higher inside a building, smoke within the building seeks an outlet to a lower-pressure zone outside the building.

____

____

_____ 22. Apply positive pressure at the highest point of a structure to remove smoke from multiple floors of a building.

____

____

_____ 23. Firefighters should call the building engineers to operate smoke control systems during a fire in a shopping mall.

____

____

Multiple Choice

D. Write the letter of the best answer on the blank before each statement.

_____ 1. In what structure is flashover *least* likely to occur?

a. Residential house with energy-saving glass windows
b. Industrial building with insulated steel entry doors
c. Office building completely equipped with vapor barriers
d. Grocery store with laminated glass doors and windows

_____ 2. What is the best way for firefighters to locate areas where extra insulation has been added to existing roofs and attic areas?

a. Ask the tenant of the building.
b. Perform a pre-incident survey.
c. Note the areas during forcible entry.
d. Mark the sites during the primary search.

____ 3. What causes heat, smoke, and fire gases to travel upward to the highest point in an area until they are trapped by a roof or ceiling?
   a. Convection
   b. Ventilation
   c. Insulation
   d. Exclusion

____ 4. What ventilation method should be used to prevent a backdraft when backdraft conditions have developed?
   a. Mechanical
   b. Natural
   c. Vertical
   d. Horizontal

____ 5. Which of the following signs does *not* indicate impending backdraft?
   a. Smoke-stained windows
   b. Black smoke rolling continually from the building
   c. Pressurized smoke coming from small cracks
   d. Little visible flame coming from the exterior of the building

____ 6. The density of smoke is in direct ratio to the ____.
   a. Quantity of fire loading
   b. Abundance of carbon monoxide
   c. Size of the fire
   d. Amount of suspended particles

____ 7. What are the *initial* factors to consider in determining whether to use horizontal or vertical ventilation?
   a. Building type and design
   b. Number and size of wall openings
   c. Availability and involvement of fire escapes
   d. Number of windows and roof openings

____ 8. How many more personnel are required for ventilating a high-rise than are required for an average residential building?
   a. One and a half times as many
   b. Two to three times as many
   c. Four to six times as many
   d. Ten times as many

____ 9. What rule of thumb should firefighters use to decide where to open a roof?
   a. Open as near to the escape route as possible.
   b. Open as directly over the fire as possible.
   c. Open as close to the middle as possible.
   d. Open as far from the flames as possible.

____ 10. If wind direction permits, where should a firefighter enter a ventilated building for fire extinguishment?

a. With the flow of the ventilation
b. Against the flow of the ventilation
c. As near the fire as possible
d. As far from the fire as possible

____ 11. With whom should the roof team be in constant communication?

a. Interior rescue crew leader
b. Exterior fire suppression crew leader
c. Building engineer
d. Incident commander

____ 12. Who is responsible for ensuring that only the required openings are made in a roof?

a. Roof crew leader
b. Interior crew leader
c. Exterior crew leader
d. Incident commander

____ 13. Where should a roof ladder be secured before firefighters operate from it?

a. On the roof ledge
b. Over the roof peak
c. Against an extension ladder
d. To a solid roof structure (e.g., chimney)

____ 14. What safety procedure is recommended for power tools used for roof ventilation?

a. Start power tools on the ground to test for operation.
b. Use extreme caution when carrying running power tools up a ladder.
c. Hoist, rather than carry, running power tools to personnel on a roof.
d. Avoid using power tools for roof ventilation.

____ 15. How far above the roof line should a ground ladder extend?

a. 2 rungs
b. 5 rungs
c. 7 rungs
d. 9 rungs

____ 16. Where should firefighters place elevated platforms for roof ventilation?

a. 2 feet *(0.6 m)* below the roof level
b. 1 foot *(0.3 m)* below the roof level
c. Even with the roof level
d. 1 foot *(0.3 m)* above the roof level

____ 17. What condition indicates that a roof is unsafe?
a. Small flames coming from the windows
b. Heavy black smoke coming from the windows
c. Springy- or spongy-feeling roof
d. Visible flames inside the structure

____ 18. For what skylight material type should firefighters use conventional forced entry, not frame removal?
a. Plexiglas® acrylic plastic
b. Lexan® plastic
c. Wired glass
d. Shatter-type glass

____ 19. What shape and size opening should a firefighter commonly make for roof ventilation?
a. Square; at least 4 × 4 feet *(1.2 m by 1.2 m)*
b. Triangle; roughly equal sides of 4 feet *(1.2 m)* each
c. Circle; about 3 feet *(1 m)* in diameter
d. Rectangle; no more than 3 feet *(1 m)* wide and 4 feet *(1.2 m)* long

____ 20. Which roof type is elevated in the center and thus forms a slope to the edges?
a. Arched
b. Pitched
c. Tilted
d. Angled

____ 21. What roof type may be constructed with bowstring trusses as supporting members?
a. Arched
b. Pitched
c. Tilted
d. Angled

____ 22. What concealed, unvented spaces of bowstring roofs can create dangerous ventilation problems and contribute to the spread of fire?
a. Attics
b. Crawl spaces
c. Cocklofts
d. Air ducts

____ 23. What is the safest way for firefighters to ventilate an arched roof?
a. Working from roof ladders in groups of two
b. Working from aerial ladders or platforms
c. Standing on extension ladders secured by harnesses
d. Sitting on foam-grip pads to prevent slipping

____ 24. For what type building is light-gauge, cold-formed steel usually used?
a. Apartment complex
b. Hospital
c. Grocery store
d. Industrial building

____ 25. What is true about trench ventilation?

a. The trench must be at least 3 feet *(1 m)* wide.
b. The trench should be cut directly behind the advancing fire.
c. The trench must extend from one exterior wall to the opposite exterior wall.
d. The trench should be cut directly above the fire to adequately remove heated smoke and gases.

____ 26. What construction type creates the hazard of a basement fire extending directly to the attic?

a. Balloon-frame
b. A-frame
c. Modular
d. Geodesic

____ 27. What basement ventilation method should be used *only* as a last resort?

a. Forcing belowground-level windows in wells
b. Using stairwell or hoistway shafts
c. Opening ground-level windows
d. Cutting the floor near a ground-level door

____ 28. Why do firefighters use elevated streams at a building fire?

a. To reduce the thermal column within the building
b. To lessen sparks and flying embers coming from a building
c. To disrupt the movement of fire gases exiting the building
d. To force superheated air and gases downward in the building

____ 29. Which of the following scenarios will *not* upset the normal processes of thermal layering and horizontal ventilation?

a. Opening a door or window on the leeward side of a structure prior to opening a door on the windward side
b. Opening a door or window on the windward side of a structure prior to opening a door on the leeward side
c. Opening doors and windows between the advancing fire fighting crews and the established ventilation exit point
d. Standing in a doorway through which an established air current is moving

____ 30. What is a disadvantage of using forced ventilation?

a. Increases smoke damage
b. Allows less positive control
c. Requires special equipment
d. Slows the removal of contaminants

____ 31. How can firefighters prevent the churning action of recirculating air around ventilation fans?

a. Leave the area around the fan open.
b. Cover the area around the fan with salvage covers.
c. Move the fans closer to the doorway or opening.
d. Place the fans diagonally in the doorway or opening.

____ 32. By what percentage does a window screen cut the effectiveness of airflow during negative-pressure ventilation?

a. 20
b. 30
c. 40
d. 50

____ 33. What exterior openings should remain open during a positive-pressure operation?

a. At least one door or window in addition to the opening where the smoke is leaving the building
b. One door or window on each side of the building
c. One door or window adjacent to the opening where the smoke is leaving the building
d. Only the door or window where the smoke is leaving the building

____ 34. How can firefighters speed up the process of removing smoke from a building during positive-pressure ventilation?

a. Placing additional blowers at the entry point
b. Leaving all interior doors open
c. Decreasing the velocity of air movement
d. Applying positive pressure at the highest level of the building

____ 35. What is a benefit of positive-pressure ventilation?

a. Works well on heavily damaged structures
b. Decreases interior carbon monoxide levels
c. Allows for fan placement that does not interfere with ingress or egress
d. Aids in detecting and extinguishing hidden fires

____ 36. Which of the following statements is true about hydraulic ventilation?

a. The larger the ventilation opening, the faster the ventilation process will go.
b. The fog pattern should cover 75 to 80 percent of the window or door opening from which the smoke will be pushed out.
c. The nozzle tip should be at least 3 feet *(1 m)* back from the ventilation opening.
d. Hydraulic ventilation combines the use of natural openings and wind direction with fan placement.

____ 37. When should firefighters identify smoke control systems?
  a. During the primary search
  b. During pre-incident planning
  c. After the ventilation process
  d. During the fire scene size-up

____ 38. What is *HVAC?*
  a. Horizontal ventilation and air control
  b. Heating, ventilation, and air-conditioning
  c. Horizontal and vertical access check
  d. High-voltage alternating current

Identify

E. Identify the following roof types. Write the correct name below each.

______________________ 1.

______________________ 2.

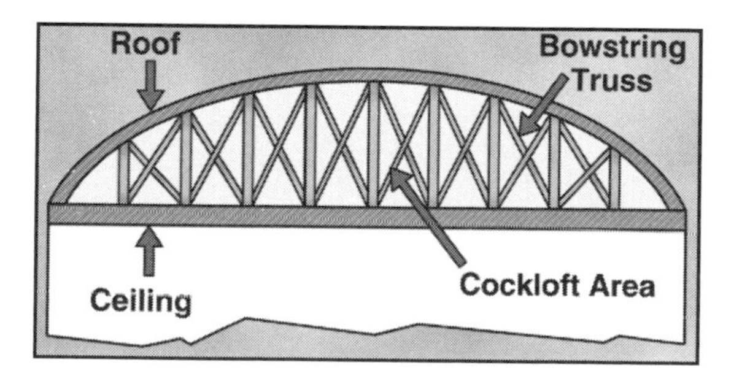

______________________ 3.

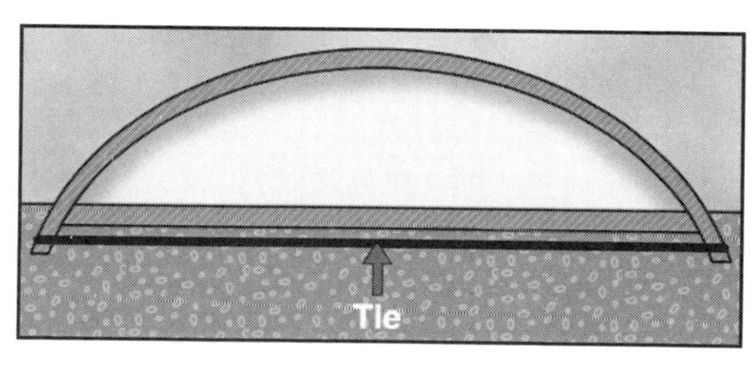

______________________ 4.

________________________________ 5.

________________________________ 6.

Chapter 11
Water Supply

Chapter 11 Water Supply

FIREFIGHTER I

Matching

A. Match to their definitions terms associated with water supply. Write the appropriate letters on the blanks.

____ 1. Water or liquid suitable for drinking

____ 2. The process of raising water from a static source to supply a pumper

____ 3. A fire hydrant that receives water from only one direction

____ 4. Water flow to a hydrant from two or more directions

____ 5. A water distribution system that provides circulating feed from several mains

____ 6. A fire hydrant used in climates where freezing weather is expected

____ 7. A fire hydrant with either a compression-type valve at each outlet or one valve in the bonnet that controls the flow of water to all outlets; entire hydrant always filled with water

____ 8. The hauling of water from a supply source to portable tanks from which water may be drawn to fight a fire

a. Grid system
b. Drafting
c. Potable
d. Dry-barrel hydrant
e. Dead-end hydrant
f. Wet-barrel hydrant
g. Circulating feed
h. Water shuttling
i. Grade arrangement

B. **Match valve types to their descriptions. Write the appropriate letters on the blanks.**

____ 1. Visually shows whether the gate or valve seat is open, closed, or partially closed

____ 2. Consists of a hollow metal post attached to the valve housing

____ 3. Has a yoke on the outside with a threaded stem that controls the gate's opening or closing

____ 4. Is usually the nonrising stem type; as the valve nut is turned by the valve key, the valve either rises or lowers to control the water flow

____ 5. Is tight closing and usually has a rubber or a rubber-composition seat that is bonded to the valve body

a. Gate
b. Indicating
c. Post indicator
d. Nonindicating
e. OS&Y
f. Butterfly

True/False

C. **Write *True* or *False* before each of the following statements. Correct those statements that are false.**

________ 1. Fire departments rely on the water department to determine locations for and types of fire hydrants.

____

____

________ 2. In small towns, domestic/industrial water requirements far exceed water requirements for fire protection.

____

____

________ 3. Direct pumping systems use one or more pumps that take water from the primary source and discharge it through the filtration and treatment processes.

____

____

_____ 4. Most communities no longer use the gravity system.

_____ 5. Water pressure is reduced as water flows through pipes.

_____ 6. Two or more primary feeders should run by separate routes from the water supply source to high-risk and industrial districts of a community.

_____ 7. Secondary feeders should run parallel to primary feeders.

_____ 8. Valves in water distribution systems provide minimal control over the flow of water through the piping.

_____ 9. A well-run water utility places aboveground markers at each valve location.

_____ 10. Most private fire protection system valves are the indicating type.

_____ 11. Nonindicating valves in water distribution systems are normally aboveground.

_______ 12. High friction loss may occur if valves are left only partially open.

____

____

_______ 13. Valve impairments may go undetected until a fire occurs or until detailed inspections and fire flow tests are made.

____

____

_______ 14. Hard suction lines do not require strainers for drafting from a natural source.

____

____

_______ 15. Commercial tank-connecting devices are generally more efficient than jet siphons for maintaining portable tank water levels.

____

____

_______ 16. Before opening a portable tank, a heavy tarp should be spread on the ground to help protect the tank's liner.

____

____

_______ 17. Gravity dumps may be activated only by the apparatus driver/operator.

____

____

_______ 18. During relay pumping, the apparatus with the greatest pumping capacity should be located at the fireground.

____

____

Multiple Choice

D. Write the letter of the best answer on the blank before each statement.

____ 1. Who should be considered the experts on water supply problems?
- a. Fire department officers
- b. Water department officials
- c. City planning administrators
- d. Fire department inspectors

____ 2. What is an example of a groundwater supply?
- a. Water well
- b. Lake
- c. River
- d. Pond

____ 3. What water needs must be included in an engineering estimate?
- a. Commercial, industrial, and domestic
- b. Industrial, domestic, and drought reserve
- c. Domestic, drought reserve, and commercial
- d. Fire fighting, domestic, and industrial

____ 4. Which of the following systems is *not* a method for moving water?
- a. Direct pumping
- b. Circulating
- c. Gravity
- d. Combination

____ 5. Where should elevated tanks for a gravity system be located?
- a. No more than a few hundred feet *(meters)* from the nearest water supply source in the distribution area
- b. No less than 300 feet *(90 m)* above sea level
- c. At least several hundred feet *(meters)* above the highest point in the distribution system
- d. At approximately every 10 square-mile *(26 km²)* section within the water distribution system

____ 6. What is another name for circulating feed?
- a. Looped line
- b. Network pipe
- c. Combination grid
- d. Grid system

____ 7. What part of a grid system consists of large pipes that convey large quantities of water to various points of the system?
- a. Direct feeders
- b. Primary feeders
- c. Secondary feeders
- d. Distributors

____ 8. What part of the grid system is comprised of small mains serving individual fire hydrants and consumer blocks?

a. Direct feeders
b. Primary feeders
c. Secondary feeders
d. Distributors

____ 9. What part of the grid system is a network of intermediate-sized pipes that reinforce the grid within the various loops of the main lines?

a. Direct feeders
b. Primary feeders
c. Secondary feeders
d. Distributors

____ 10. What is the recommended size for fire-hydrant supply mains in residential areas?

a. 5 inches *(125 mm)*
b. 6 inches *(150 mm)*
c. 7 inches *(180 mm)*
d. 8 inches *(200 mm)*

____ 11. Residential fire-hydrant supply mains should be closely gridded by ____ cross-connecting mains at intervals of not more than ____.

a. 6-inch *(150 mm)*, 400 feet *(120 m)*
b. 7-inch *(180 mm)*, 500 feet *(150 m)*
c. 8-inch *(200 mm)*, 600 feet *(180 m)*
d. 9-inch *(230 mm)*, 700 feet *(210 m)*

____ 12. In business and industrial districts, the minimum recommended size for fire-hydrant supply mains is a(n) ____ main with cross-connecting mains every ____.

a. 8-inch *(200 mm)*, 600 feet *(180 m)*
b. 10-inch *(250 mm)*, 800 feet *(240 m)*
c. 12-inch *(300 mm)*, 1,000 feet *(300 m)*
d. 14-inch *(350 mm)*, 1,200 feet *(360 m)*

____ 13. What is the recommended size for fire-hydrant supply mains used on principal streets and in long mains *not* cross-connected at frequent intervals?

a. 8-inch *(200 mm)*
b. 10-inch *(250 mm)*
c. 12-inch *(300 mm)*
d. 14-inch *(350 mm)*

____ 14. How often should water valves be operated to keep them in good condition?

a. Twice per year
b. Once per year
c. Once every two years
d. Only during fire fighting operations

____ 15. Where are the words OPEN and SHUT printed on a post indicator valve?

a. On the valve stem inside the post
b. On the yoke
c. On the valve housing
d. Beneath the threaded stem

____ 16. How far does a valve disk rotate from the fully open position to the tightly shut position?

a. Three counterclockwise turns
b. One clockwise turn
c. The number of turns marked on the valve
d. 90 degrees

____ 17. What kind of water pipes should be used in unstable or corrosive soil?

a. Steel
b. Asbestos cement
c. Ductile iron
d. Metal amalgamate

____ 18. Which of the following statements is true about dry-barrel hydrants?

a. The drain is open only when the hydrant is flowing.
b. The valve holding the water back is aboveground.
c. The barrel is filled with water only during the summer.
d. The hydrant must be completely open for the drain to close.

____ 19. What part of a hydrant is usually made of bronze?

a. Bonnet
b. Working parts
c. Footpieces
d. Barrels

____ 20. Typically, what color is a Class A hydrant?

a. Orange
b. Green
c. Light blue
d. Red

____ 21. Which hydrant class is used for water flow of 500–999 gpm *(1 900 L/min to 3 780 L/min)*?

a. AA
b. A
c. B
d. C

____ 22. How much water flow does a Class C hydrant produce?

a. Less than 500 gpm *(1 900 L/min)*
b. 500–999 gpm *(1 900 L/min to 3 780 L/min)*
c. 1,000–1,499 gpm *(3 785 L/min to 5 675 L/min)*
d. 1,500 gpm *(5 680 L/min)* or greater

____ 23. Where should the hard suction strainer be placed in a static water source?

a. On or near the bottom of the static source
b. About 10 inches *(250 mm)* below the surface of the water
c. With 24 inches *(600 mm)* of water above and below the strainer
d. Approximately 15 inches *(375 mm)* above the bottom of the static source

____ 24. From what minimum water depth can a special drafting or floating strainer draw water?

a. 1 to 2 inches *(25 mm to 50 mm)*
b. 3 to 4 inches *(75 mm to 100 mm)*
c. 5 to 6 inches *(125 mm to 150 mm)*
d. 6 to 7 inches *(175 mm to 180 mm)*

____ 25. What NFPA standard provides requirements for rural water supply operations?

a. 1500
b. 1970
c. 1971
d. 1231

____ 26. Water shuttling is recommended for distances greater than ____.

a. 500 feet *(0.2 km)*
b. 1,000 feet *(0.3 km)*
c. ¼ mile *(0.4 km)*
d. ½ mile *(0.8 km)*

____ 27. What are the keys to efficient water shuttling operations?

a. Ample tankers/tenders and adequate personnel
b. Large portable tanks and large water supply
c. Fast-fill and fast-dump times
d. Sufficient hard suction lines and strainers

____ 28. Where should water supply officers be positioned during water shuttling activities?

a. At the fill site only
b. In the tanker/tender vehicles
c. At the incident command post
d. At both the dump and fill sites

____ 29. What device permits use of most of the water in the portable reservoir during water shuttling operations?

a. Low-level intake device
b. Drafting/floating strainer
c. Hard suction line
d. Jet siphon

____ 30. What is the water capacity range for portable tanks?

a. 500 gallons *(2 000 L)* and upward
b. 750 gallons *(3 000 L)* and upward
c. 1,000 gallons *(4 000 L)* and upward
d. 1,500 gallons *(6 000 L)* and upward

____ 31. What device maintains the water level in one portable tank for the pumper, while water tankers/tenders dump water into another during water shuttling operations?

a. Low-level intake device
b. Drafting/floating strainer
c. Hard suction line
d. Jet siphon

____ 32. Who should pump the water from the tanker/tender during water shuttling operations?

a. Firefighter
b. Driver/operator
c. Water supply officer
d. Incident commander

____ 33. Who should be appointed to determine the appropriate distance between pumpers for relay pumping and to coordinate water supply operations?

a. Firefighter
b. Driver/operator
c. Water supply officer
d. Incident commander

Identify

E. Identify the following abbreviations associated with water supply. Write the correct interpretation before each.

______________________________ 1. PIV

______________________________ 2. OS&Y

F. Identify types of water main valves. Write the correct name below each.

______________________ 1.

______________________ 2.

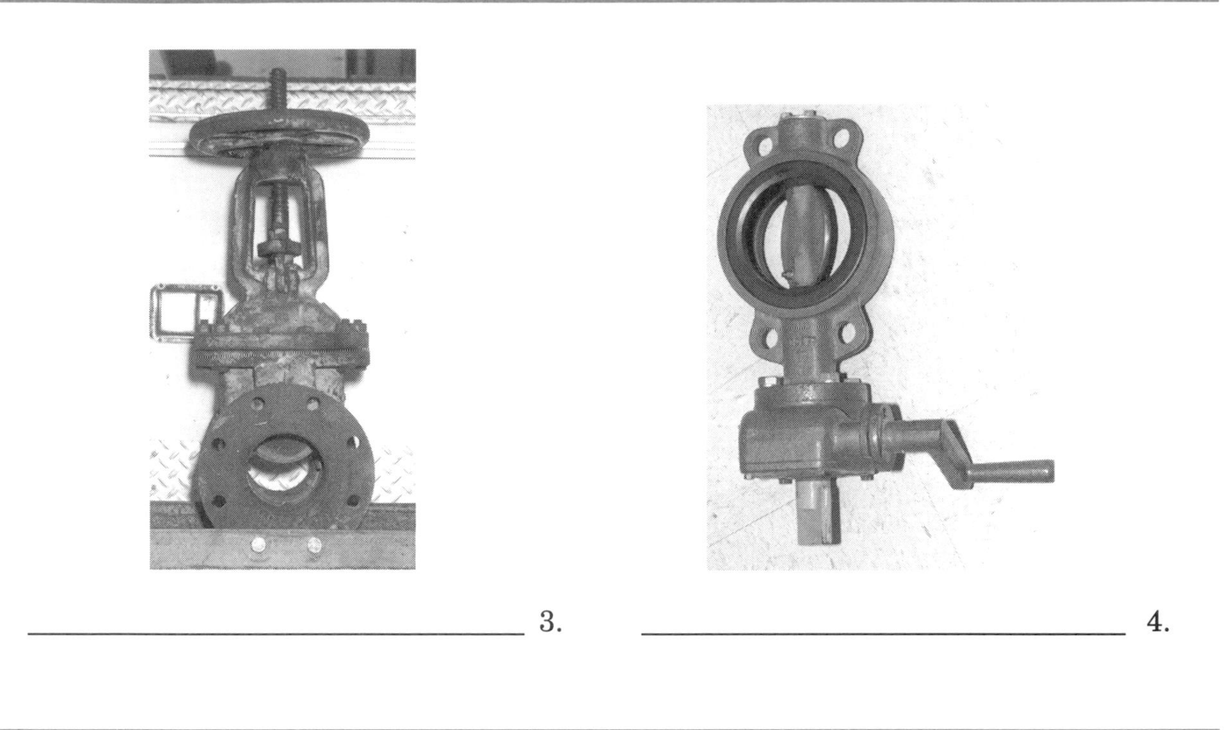

_______________________ 3. _______________________ 4.

Label

G. Label the parts of the water distribution system illustrated below.

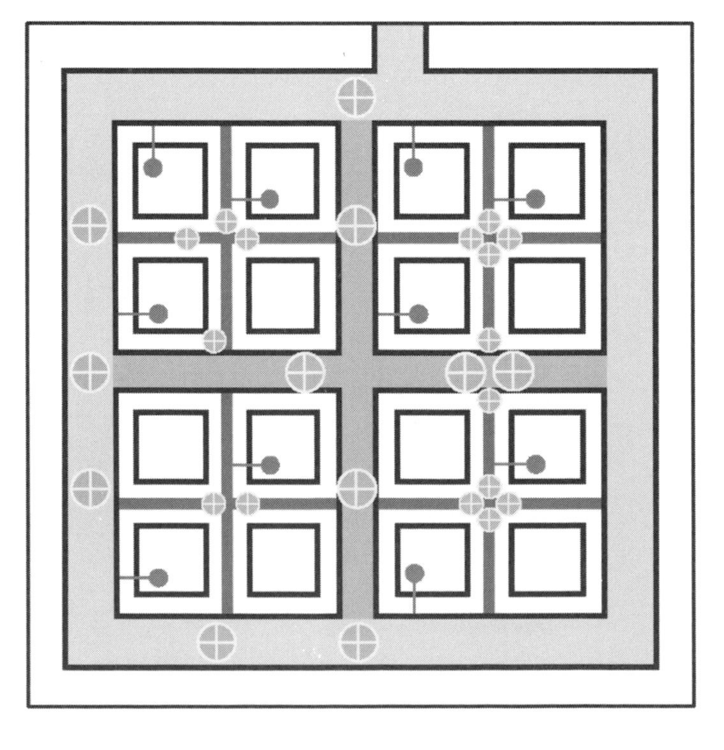

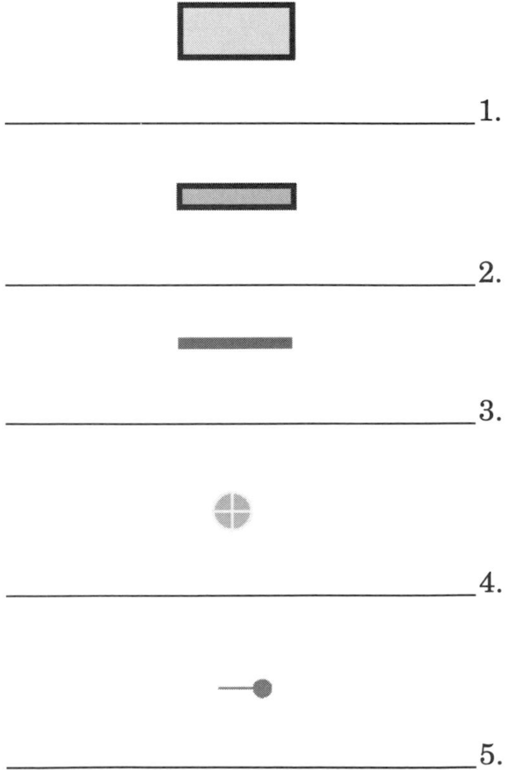

_______________________ 1.

_______________________ 2.

_______________________ 3.

_______________________ 4.

_______________________ 5.

FIREFIGHTER II

Matching

A. Match types of water supply pressure to their definitions. Write the appropriate letters on the blanks.

____ 1. The forward velocity of water in a conduit (either pipe or hose) of a certain size

____ 2. The stored potential energy that is available to force water through pipe, fittings, fire hose, and adapters

____ 3. The pressure found in water distribution systems during periods of ordinary consumption demand

____ 4. The part of the total available pressure that is not used to overcome friction or gravity while forcing water through pipe, fittings, fire hose, and adapters

____ 5. The forward velocity pressure at a discharge opening, either at a hydrant discharge or a nozzle discharge orifice, while water is flowing

a. Pressure
b. Mainline pressure
c. Normal operating pressure
d. Residual pressure
e. Static pressure
f. Flow pressure

True/False

B. Write *True* or *False* before each of the following statements. Correct those statements that are false.

________ 1. Fluoride and oxygen are often added to water at processing or treatment facilities.

____

____

________ 2. In the fire service, water pressure is measured in pounds per square foot or kilograms per square meter.

____

____

_____ 3. During pre-incident planning, firefighters *cannot* determine the possible effect that weather may have on the amount of water available for fire fighting operations.

____

____

Multiple Choice

C. Write the letter of the best answer on the blank before each statement.

_____ 1. Which of the following processes is *not* used to remove contaminants from water?

a. Sedimentation
b. Coagulation
c. Adding bacteria
d. Adding fluoride

_____ 2. What is the fire department's main concern regarding treatment facilities?

a. Inadequate filtration systems that may leave harmful contaminants in water used for fire fighting operations
b. Maintenance errors or natural disasters that could cause a reduction in the volume and pressure of water
c. Insufficient treatment facility personnel to properly test all of the district's fire hydrants
d. Inability of the water treatment facility to properly place hydrants and water lines needed for fire fighting operations

_____ 3. What test result indicates that a dry-barrel hydrant is draining properly?

a. After flowing the hydrant, then capping all but one discharge, the tester should be able to feel a slight vacuum over the open discharge.
b. After shutting off the water supply main, then uncapping two of the discharges, the tester should not be able to see any water in the barrel.
c. After flowing the hydrant for several minutes, then capping all of the discharges and waiting several minutes, the tester should hear a sharp ringing sound when the hydrant is tapped with a metal rod.
d. Before flowing the hydrant, the tester should not be able to see any erosion or water accumulation around the base of the hydrant.

_____ 4. Who should determine the location, spacing, and distribution of fire hydrants?

a. City planners
b. Fire chief or fire marshal
c. Water department personnel
d. Water supply officer

____ 5. Generally, what is the maximum distance that fire hydrants should be spaced in high-value districts?

a. 200 feet *(60 m)*
b. 300 feet *(90 m)*
c. 400 feet *(120 m)*
d. 500 feet *(150 m)*

____ 6. Intermediate hydrants should be placed when distances between intersections exceed ____.

a. 750 to 800 feet *(225 m to 240 m)*
b. 550 to 600 feet *(165 m to 180 m)*
c. 350 to 400 feet *(105 m to 120 m)*
d. 200 to 250 feet *(60 m to 75 m)*

____ 7. What tool do firefighters use to measure the flow pressure coming from a hydrant?

a. Velocity recorder
b. Air pressure gauge
c. Petcock gauge
d. Pitot tube

____ 8. When should firefighters attempt to identify, mark, and record alternative water supply sources?

a. During pre-incident planning
b. During initial size-up
c. After initiating the fire attack
d. As soon as the water supply officer arrives

Identify

D. Identify the following abbreviations associated with water supply. Write the correct interpretation before each.

______________________________ 1. psi

______________________________ 2. kPa

Chapter 12
Fire Hose

Chapter 12 Fire Hose

FIREFIGHTER I

Matching

A. Match hose types to their uses. Write the appropriate letters on the blanks.

____ 1. Length coupled with another length to produce a continuous hoseline

____ 2. Length used to connect a fire department pumper or a portable pump to a nearby water source

____ 3. Length used to transfer water from a pressurized water source, such as a hydrant, to the pump intake

____ 4. Length used primarily to draft water from an open water source

____ 5. Length used between the water source and the attack pumper to provide large volumes of water

____ 6. Length between the attack pumper and the nozzle used to control and extinguish fire

a. Soft sleeve
b. Attack
c. Intake
d. Stock
e. Supply
f. Hard suction (hard sleeve)
g. Section

B. Match to their definitions terms associated with fire hose. Write the appropriate letters on the blanks. Definitions are continued and terms are repeated on next page.

____ 1. The force created by rapid deceleration of water

____ 2. The portion of a coupling that serves as the point of attachment to the hose

____ 3. Special tool that fits against coupling lugs

____ 4. Special type of thread design in which the beginning of the thread is "cut" to provide a positive connection between the first threads of opposing couplings

a. Spanner
b. Male coupling
c. Water hammer
d. Expansion-ring gasket
e. Swivel gasket
f. Shank
g. Higbee cut
h. Higbee indicator
i. Finish load
j. Hose bed
k. Minuteman load

____ 5. The shallow indention scalloped into one of the rocker lugs on the swivel of a coupling

____ 6. Device used at the end of the hose where the hose is expanded into the shank of the coupling

____ 7. Device used to make a connection watertight when threaded couplings are connected

____ 8. Hose pulled and advanced by one person

____ 9. Compartment used for storing fire hose

____ 10. Arrangement of hose usually placed on top of a hose load and connected to the end of the load

a. Spanner
b. Male coupling
c. Water hammer
d. Expansion-ring gasket
e. Swivel gasket
f. Shank
g. Higbee cut
h. Higbee indicator
i. Finish load
j. Hose bed
k. Minuteman load

True/False

C. Write *True* or *False* before each of the following statements. Correct those statements that are false.

________ 1. References made to the diameter of fire hose refer to the dimensions of the outside diameter of the hose.

____

____

________ 2. Usually little can be done at fires to protect hose from injury.

____

____

________ 3. Exposing hose to intense heat can weaken the fabric covering and dry the rubber lining.

____

____

________ 4. Unlike wet hose, dry hose is not affected by freezing temperatures.

____

____

_______ 5. The most commonly used couplings are the quarter turn and oilfield rocker lug.

____

____

_______ 6. Drop-forged couplings are weak and are rarely used.

____

____

_______ 7. Only male couplings have lugs on the shank.

____

____

_______ 8. Pin lugs help the coupling slide over obstructions when hose is moved on the ground or around objects.

____

____

_______ 9. Storz-type couplings are designed to be connected and disconnected with only two-thirds of a turn.

____

____

_______ 10. All parts of a fire hose coupling are susceptible to damage.

____

____

_______ 11. Firefighters should use hose-washing machines to clean hose coupling swivels that have become stiff or sluggish from dirt or other foreign matter.

____

____

_______ 12. The swivel gasket and the expansion-ring gasket are interchangeable.

____

____

_______ 13. A damaged coupling or piece of hose is often denoted by straight rolling a hose so that the male coupling is exposed.

____

____

_______ 14. Fire departments should develop specific techniques for coupling and uncoupling hose to increase speed and accuracy during emergency conditions.

____

____

_______ 15. Most hose beds have solid bases.

____

____

_______ 16. A split hose bed allows the apparatus to have both forward and reverse lays.

____

____

_______ 17. The horseshoe load works well for large diameter hose.

____

____

_______ 18. Of the three different supply hose loads, the flat load is the easiest to load.

____

____

_______ 19. Large diameter hose should not be flat-loaded from the ground.

____

____

_______ 20. Finishes for forward lays are usually elaborate and not designed for speed.

____

____

_________ 21. The minuteman load is particularly well suited for a narrow bed.

_________ 22. Noncollapsible hose should be loaded onto booster hose reels one layer at a time in an even manner.

_________ 23. Threaded-coupling supply hose is usually arranged in the hose bed so that when hose is laid, the end with the male coupling is toward the water source.

_________ 24. Hose beds set up for forward lays should be loaded so that the male coupling is first off the hose bed.

_________ 25. Typically, the four-way hydrant is preconnected to the end of the supply line.

_________ 26. Laying hose from the hydrant to the fire is called a reverse lay.

_________ 27. Laying hose from the incident scene back to the water source has become the standard method for setting up a relay pumping operation when using large diameter hose as a supply line.

_________ 28. A pumper must be placed at the water source when drafting.

_______ 29. Firefighters should use a four-way hydrant valve when reverse-laying a supply hose.

____

____

_______ 30. A soft sleeve hose should be used when drafting from a static water supply source.

____

____

_______ 31. Not all hydrants have large steamer discharges capable of accepting direct connections from soft sleeve hoses.

____

____

_______ 32. Making connections with hard suction hose is considerably more difficult than making connections with soft intake hose.

____

____

_______ 33. The working line drag is a slow method of moving fire hose because it is accomplished by only one firefighter.

____

____

_______ 34. If it becomes necessary to advance a charged hoseline up a stairway, firefighters should clamp it first.

____

____

_______ 35. Hoselines should *not* typically be charged before advancing them down a stairway.

____

____

_______ 36. Hoselines should be pulled from the apparatus and extended to the fire area to fight fire on an upper floor of a tall building.

____

____

_______ 37. Once a hoseline has been charged, it should not be drained before advancing it up a ladder.

____

____

_______ 38. Firefighters should not exceed the rated capacity of a ladder even when advancing hoselines up the ladder.

____

____

_______ 39. Firefighters can control a loose LDH by putting a kink in the hose.

____

____

_______ 40. Ideally, two or three firefighters should be used when working with a nozzle connected to a large-sized attack line.

____

____

Muliple Choice

D. Write the letter of the best answer on the blank before each statement.

____ 1. Into what lengths is fire hose most commonly cut and coupled?

a. 20 or 40 feet *(6 m or 12 m)*

b. 30 or 60 feet *(9 m or 18 m)*

c. 40 or 80 feet *(12 m or 24 m)*

d. 50 or 100 feet *(15 m or 30 m)*

____ 2. In what size range is soft sleeve hose available?

a. 1¾ to 4 inches *(45 mm to 100 mm)*

b. 2 to 5 inches *(50 mm to 125 mm)*

c. 2½ to 6 inches *(65 mm to 150 mm)*

d. 3¼ to 7 inches *(80 mm to 180 mm)*

____ 3. Which hose type is used to siphon water from one portable tank to another, usually in connection with a tanker shuttle operation?

a. Soft sleeve

b. Hard suction

c. Intake

d. Attack

____ 4. What hose type is constructed of rubberized, reinforced material designed to withstand partial vacuum conditions?

a. Soft sleeve
b. Hard suction
c. Intake
d. Attack

____ 5. What NFPA standard lists specifications for fire hose?

a. 1961
b. 1963
c. 1903
d. 1970

____ 6. According to NFPA 1901, how much and what size attack hose are pumpers required to carry?

a. 400 feet *(122 m)* of 1½-, 1¾-, or 2-inch *(38 mm, 45 mm, or 50 mm)* hose
b. 300 feet *(90 m)* of 1¾-, 2-, or 2½-inch *(45 mm, 50 mm, or 65 mm)* hose
c. 200 feet *(60 m)* of 1¾-, 2-, or 2½-inch *(45 mm, 50 mm, or 65 mm)* hose
d. 175 feet *(55 m)* of 2-, 2½-, or 3-inch *(50 mm, 65 mm, or 77 mm)* hose

____ 7. How many feet of 2½-inch *(65 mm)* or larger supply hose is a pumper required to carry according to NFPA 1901?

a. 800 feet *(244 m)*
b. 1,000 feet *(305 m)*
c. 1,200 feet *(366 m)*
d. 1,600 feet *(488 m)*

____ 8. In addition to the supply hose in Question 7, what other hose lengths are pumpers required to carry according to NFPA 1901?

a. At least 20 feet *(6 m)* of large soft sleeve hose
b. Either 20 feet *(6 m)* of medium soft sleeve hose or 30 feet *(9 m)* of hard suction hose
c. At least 30 feet *(9 m)* of hard suction hose
d. Either 15 feet *(4.6 m)* of large soft sleeve or 20 feet *(6 m)* of hard suction hose

____ 9. What precaution should be taken to prevent damage to hose?

a. Slowly drive apparatus over hose.
b. Carefully fold or bend hose in the same places after each use.
c. Quickly pull hose over rough, sharp edges.
d. Slowly open and close nozzles, valves, and hydrants.

____ 10. What hose-care practice prevents thermal damage to hose?

a. Drying hose in direct sunlight
b. Laying hose on hot pavement to dry
c. Keeping the outside of woven-jacket fire hose dry
d. Allowing hose to remain in a heated area after it is dry

____ 11. Which hose type is subject to mold and mildew damage?

a. Woven-jacket
b. Rubber-jacket
c. Either woven-jacket or rubber-jacket
d. Neither woven-jacket nor rubber-jacket

____ 12. If woven-jacket hose has not been unloaded from the apparatus for ____ days, it should be removed, inspected, and reloaded.

a. 15
b. 30
c. 45
d. 60

____ 13. How often should firefighters run water through woven-jacket hose to prevent the rubber lining from drying and cracking?

a. Every 30 days
b. Every 45 days
c. Every 60 days
d. Every 90 days

____ 14. Which of the following chemicals will not damage the rubber lining in fire hose?

a. Alkalis
b. Baking soda
c. Petroleum products
d. Paint thinner

____ 15. What should be used to clean hose that has been exposed to harmful chemicals?

a. Alkalis
b. Baking soda
c. Petroleum products
d. Paint thinner

____ 16. What should be done with hose that has been exposed to hazardous materials and cannot be decontaminated?

a. Use it for practice evolutions only.
b. Keep it at the station for backup.
c. Dispose of it properly.
d. Keep it on the apparatus for backup.

____ 17. What hose type requires more cleaning than just rinsing with clear water?

a. Woven-jacket
b. Hard-rubber
c. Hard suction
d. Rubber-jacket

____ 18. What cleaning agent should be used to wash fire hose that has been exposed to oil?

a. Heavy-duty detergent
b. Mild soap or detergent
c. Hot water and bleach
d. Clear warm water

____ 19. What is the largest hose size that can be washed in the most common type of hose-washing machine?

a. 3 inches *(77 mm)*
b. 3½ inches *(90 mm)*
c. 4 inches *(100 mm)*
d. 4½ inches *(115 mm)*

____ 20. What hose type *cannot* be placed back on the apparatus while wet?

a. Hard-rubber booster
b. Hard suction
c. Rubber-jacket collapsible
d. Woven-jacket

____ 21. What materials generally make up the alloys used for fire hose couplings?

a. Steel, brass, aluminum
b. Brass, aluminum, lead
c. Brass, aluminum, magnesium
d. Magnesium, tin, brass

____ 22. What type couplings are the weakest and rarely used on modern fire hose?

a. Cast
b. Drop-forged
c. Extruded
d. Amalgamate

____ 23. What coupling type is used when the needed coupling size is smaller than the hose to which it is attached?

a. Six-piece
b. Five-piece
c. Four-piece
d. Three-piece

____ 24. Female lugs are located on what part of the coupling?

a. Shank
b. Tailpiece
c. Bowl
d. Swivel

____ 25. What type lug helps couplings slide over obstructions when a hose is moved on the ground or around objects?

a. Pin
b. Recessed
c. Rocker
d. Reel

____ 26. What coupling type is referred to as the *sexless* coupling?

a. Quarter turn
b. Threaded
c. Storz
d. Jones snap

____ 27. Which of the following procedures should *not* be followed for fire hose coupling care?

a. Leave the gasket attached for cleaning.
b. Avoid dragging couplings.
c. Examine the couplings when hose is washed and dried.
d. Remove oil from threads.

____ 28. How can a firefighter test a gasket?

a. Allow water to flow through the coupling at full pressure.
b. Pinch the gasket between the thumb and index finger.
c. Remove the gasket from the swivel and visually inspect it.
d. Send the gasket back to the manufacturer for testing.

____ 29. For which of the following situations is the straight hose roll *not* commonly used?

a. Loading hose back on the apparatus at the fire scene
b. Returning hose to quarters for washing
c. Placing hose in a rack storage
d. Deploying hose for use directly from a roll

____ 30. What is an advantage of the donut roll over the straight roll?

a. Only the male coupling is visible.
b. The female coupling is protected in the center of the roll.
c. The hose spirals easily when unrolled.
d. The hose may be quickly unrolled and placed into service.

____ 31. Which of the following hose loading guidelines should be followed, regardless of the type of hose load used?

a. The flat sides of the hose should be kept on the same plane when two sections of hose are connected.
b. The lugs on the couplings must be in alignment.
c. The couplings connecting hose sections should be tightened with a wrench.
d. Large diameter hose should be loaded with the couplings placed at the rear of the bed.

____ 32. What is the primary advantage of the horseshoe hose load?

a. It can be used for large diameter hose.
b. The hose is loaded from the ground to avoid wear on the hose edges.
c. One person can make the folds for a shoulder carry.
d. It has fewer sharp bends than flat loads.

____ 33. Which of the following is *not* true about the flat load?

a. This method is the best way to load large diameter hose.
b. The flat load is suitable for any size supply hose.
c. Hose folds do not contain sharp bends.
d. This load reduces wear from apparatus vibration during travel.

____ 34. Where should a firefighter start the hose lay for a large diameter hose?
  a. 6 to 10 inches *(150 mm to 250 mm)* from the back of the hose bed
  b. 10 to 16 inches *(250 mm to 400 mm)* from the edge of the apparatus
  c. 12 to 18 inches *(300 mm to 450 mm)* from the front of the hose bed
  d. 18 to 24 inches *(450 mm to 600 mm)* from the bottom of the hose rack

____ 35. Which statement is true about the reverse horseshoe finish?
  a. This finish can be used only for 2½-inch *(65 mm)* attack hose.
  b. The bottom of the *U* portion of the horseshoe is at the rear of the bed.
  c. It is made of one or two 50-foot *(15 m)* lengths of hose, each connected to one side of a wye.
  d. This finish, which uses a 2½-inch *(65 mm)* hose, also requires a 2-× 2-inch *(65 mm by 65 mm)* gated wye.

____ 36. Where should preconnected hoselines *not* be placed on apparatus?
  a. Main hose bed
  b. Longitudinal bed
  c. Side compartment
  d. Front bumper well

____ 37. What hose load is often preferred for 2- and 2½-inch *(50 mm and 65 mm)* attack lines that may be too cumbersome for other carries?
  a. Preconnected flat load
  b. Double layer load
  c. Minuteman load
  d. Triple layer load

____ 38. For what type lay is the hose laid from the water source to the fire?
  a. Forward lay
  b. Reverse lay
  c. Split lay
  d. Combination lay

____ 39. For which lay is the primary advantage that the pumper can remain at the incident scene so that its hose, equipment, and tools can be quickly obtained if needed?
  a. Forward lay
  b. Reverse lay
  c. Split lay
  d. Combination lay

____ 40. What is a disadvantage of the forward lay?
  a. The pumper must remain at the hydrant throughout the fire attack.
  b. An additional pumper may have to boost the pressure in the line.
  c. The initial attack is delayed while hose is removed at the fire location before the pumper proceeds to the water source.
  d. The pump operator must stay with the pumper at the water source.

____ 41. The person catching the hydrant must have all of the following tools, *except* a ____.
  a. Spanner wrench
  b. Hydrant wrench
  c. Gated wye
  d. Four-way hydrant valve

____ 42. Generally, how much hose should be manually pulled from the apparatus before starting a hose lay?

a. 3 feet *(1 m)*
b. 9 feet *(2.7 m)*
c. 15 feet *(5 m)*
d. 30 feet *(9 m)*

____ 43. What is the best way to anchor a hose at the location from which the lay is being made?

a. Place the hose behind the tire of an apparatus and slowly back onto it.
b. Have two firefighters hold the hose until the lay is complete.
c. Wrap the end of the hose around a stationary object.
d. Allow enough slack so that the hose does not need to be anchored.

____ 44. Which lay is the most direct way to supplement hydrant pressure and perform drafting operations?

a. Forward lay
b. Reverse lay
c. Split lay
d. Combination lay

____ 45. When the first pumper to arrive at the scene must work alone for an extended period of time, how is reverse-laid hose used?

a. For drafting
b. For supplementing hydrant pressure
c. As a water supply line
d. As an attack line

____ 46. What is hard sleeve hose used for if it is marked FOR VACUUM USE ONLY?

a. Drafting
b. Supplementing hydrant pressure
c. Water supply line
d. Attack line

____ 47. For what hose group is the direction of the hose lay inconsequential?

a. Hose used for drafting operations
b. Preconnected water supply hose
c. Hose equipped with Storz couplings
d. Horseshoe-loaded hose

____ 48. What hose advancement method simply involves pulling the hose from the compartment and walking toward the fire?

a. Minuteman load
b. Preconnected flat load
c. Triple layer load
d. Wyed lines

____ 49. What hose advancement method involves placing the nozzle and the fold of the first tier on the shoulder and walking away from the apparatus?

a. Minuteman load
b. Preconnected flat load
c. Triple layer load
d. Wyed lines

____ 50. Which of the following hose loads is deployed without dragging any of the hose on the ground?

a. Minuteman load
b. Preconnected flat load
c. Triple layer load
d. Wyed lines

____ 51. When removing hose from the apparatus, what hose load must be put on the shoulder one section at a time?

a. Horseshoe
b. Accordion
c. Wyed
d. Working line

____ 52. Which of the following is a safety guideline for advancing hose into a burning structure?

a. Place the backup firefighter(s) on the opposite side of the line from the firefighter at the nozzle.
b. Before entering the fire area, bleed air from the hoseline once it is charged.
c. Open the nozzle before opening the door to safeguard against possible backdraft or flashover conditions.
d. Use ventilation openings to advance hose into the structure.

____ 53. For what size hose should a gated wye be used when connecting to a standpipe?

a. 2-inch *(50 mm)*
b. 2½-inch *(65 mm)*
c. 3-inch *(77 mm)*
d. 3½-inch *(90 mm)*

____ 54. What is the *least* desirable way to get hose to an elevated position?

a. Carry hose up a stairway in a bundle.
b. Hoist hose up to a window or landing using a rope.
c. Advance unbundled hose up a stairway.
d. Advance charged hoselines up a ladder.

____ 55. How should firefighters space themselves when advancing an *uncharged* hoseline up a ladder?

a. With no more than 20 rungs between them
b. So that the firefighters are within reach of each other
c. With no more than one firefighter per fly section
d. With no less than two firefighters per fly section

____ 56. How should firefighters space themselves when advancing a *charged* hoseline up a ladder?

a. With no more than 20 rungs between them
b. So that the firefighters are within reach of each other
c. With no more than one firefighter per fly section
d. With no less than two firefighters per fly section

____ 57. What is the safest way to control a loose line?

a. Close the valve at the pump or hydrant.
b. Position a hose clamp at some point in the hoseline.
c. Put a kink in the hose at a point near the break.
d. Drive apparatus over the hose and stop with one tire on the hose.

____ 58. How much hose is required to replace a burst section of hose?

a. One section of hose
b. 5 feet *(1.5 m)* more than the damaged section
c. 10 feet *(3 m)* more than the damaged section
d. Two sections of hose

____ 59. How far straight behind the firefighter should the hoseline be laid during the one-firefighter method of operating a medium-size hose and nozzle?

a. 3 feet *(1 m)*
b. 5 feet *(1.5 m)*
c. 7 feet *(2.1 m)*
d. 10 feet *(3 m)*

____ 60. Where should the backup firefighter be positioned for the two-firefighter method of operating a medium-size hose and nozzle?

a. On the same side of the hose and 3 feet *(1 m)* behind the nozzleperson
b. On the opposite side of the hose and 3 feet *(1 m)* behind the nozzleperson
c. On the same side of the hose and 10 feet *(3 m)* behind the nozzleperson
d. On the opposite side of the hose and 10 feet *(3 m)* behind the nozzleperson

____ 61. How can a firefighter accomplish the one-firefighter method of operating large-size attack hose in a reasonably safe manner?

a. Secure the hose to one leg with a hose strap or utility strap, and brace against a stationary object.
b. Form a large loop, cross the loop over the line behind the nozzle, and sit where the hose crosses the direct stream.
c. Strap the hose to a stationary object approximately 3 feet *(1 m)* behind the nozzle, and maneuver the nozzle end only.
d. Reduce the water flow by overlapping a section of the hose, and then stand on that section.

____ 62. What is the most mobile manner of operating large-size attack hose with the three-firefighter method?

a. All three firefighters kneel to anchor the line.
b. The nozzleperson stands and both of the backup firefighters kneel.
c. The first backup firefighter stands and the second backup firefighter kneels.
d. All three firefighters use hose straps and remain in a standing position.

Identify

E. Identify the following hose rolls. Write the correct name below each.

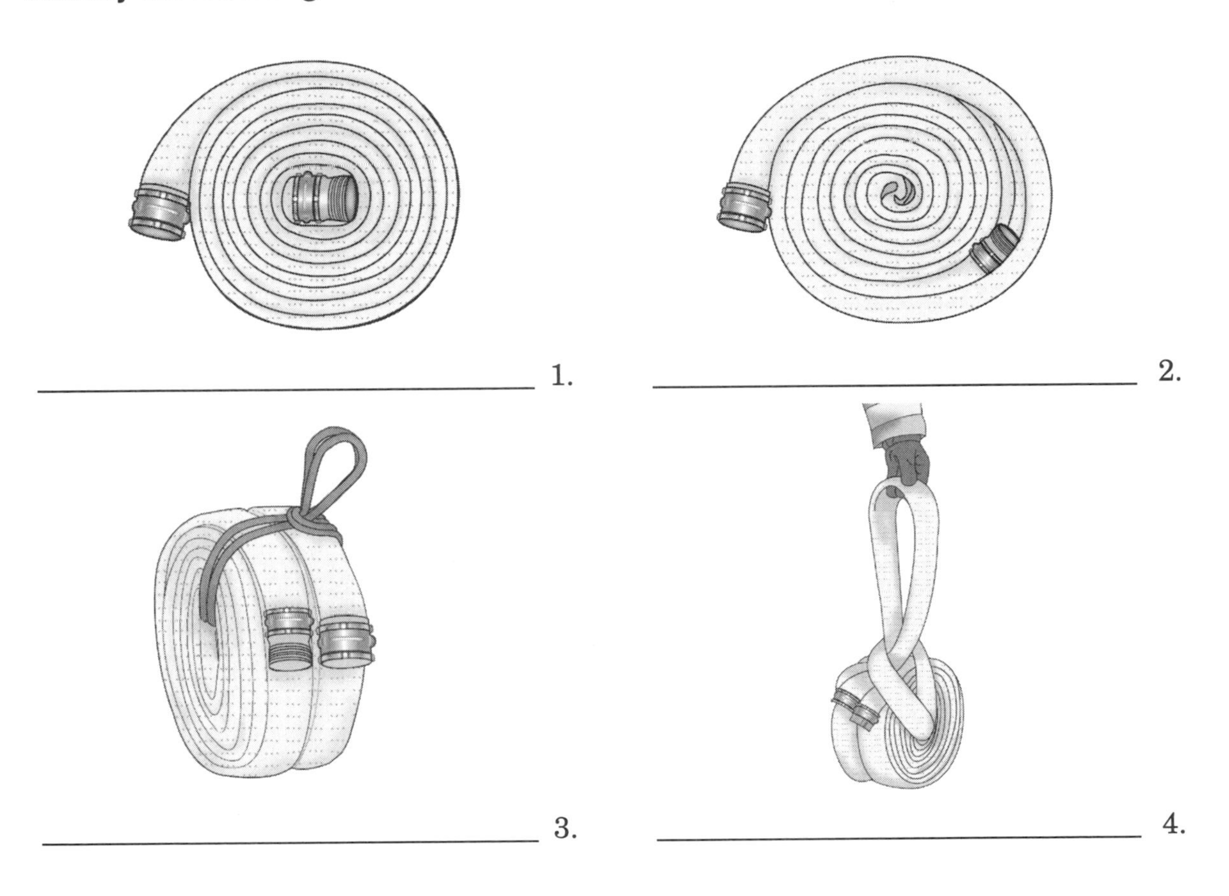

1. ______________________

2. ______________________

3. ______________________

4. ______________________

F. Identify the following hose loads and finishes. Write the correct name below each.

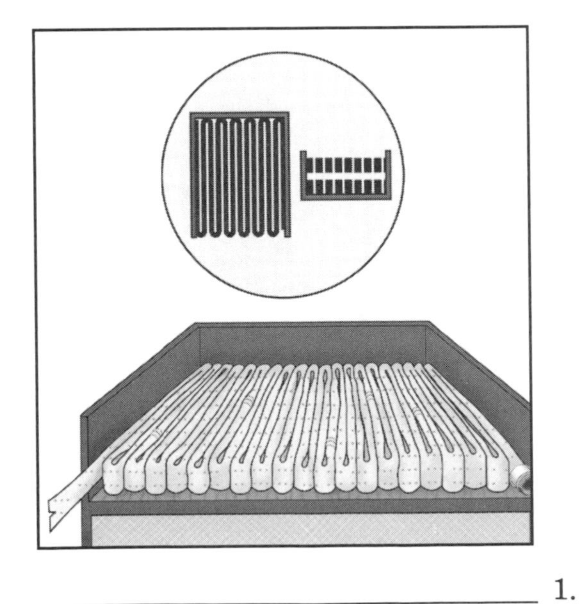

1. ______________________

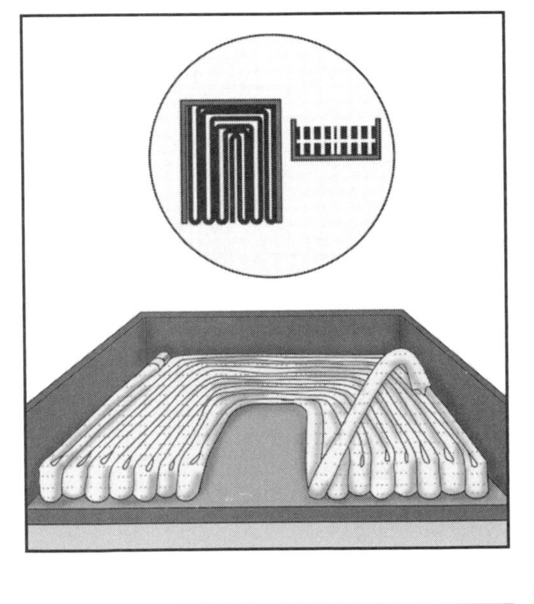

2. ______________________

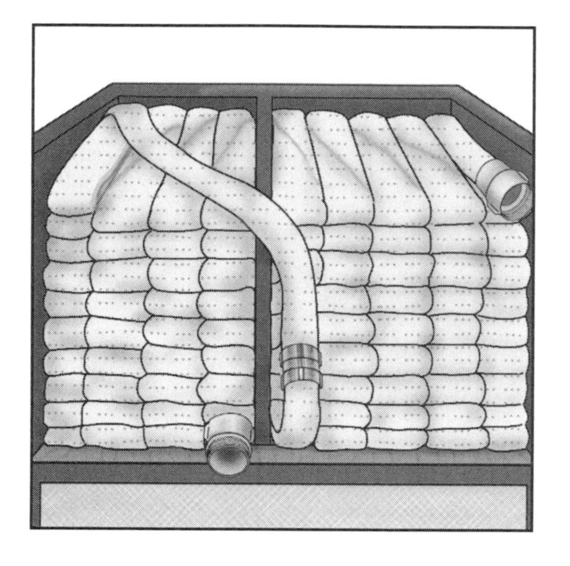

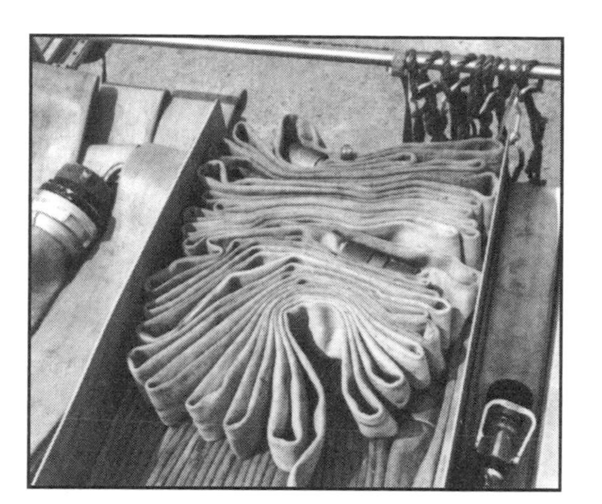

_______________ 3.

_______________ 4.

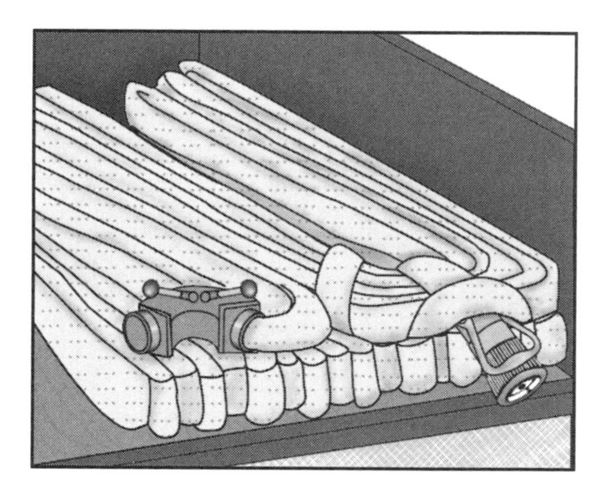

_______________ 5.

_______________ 6.

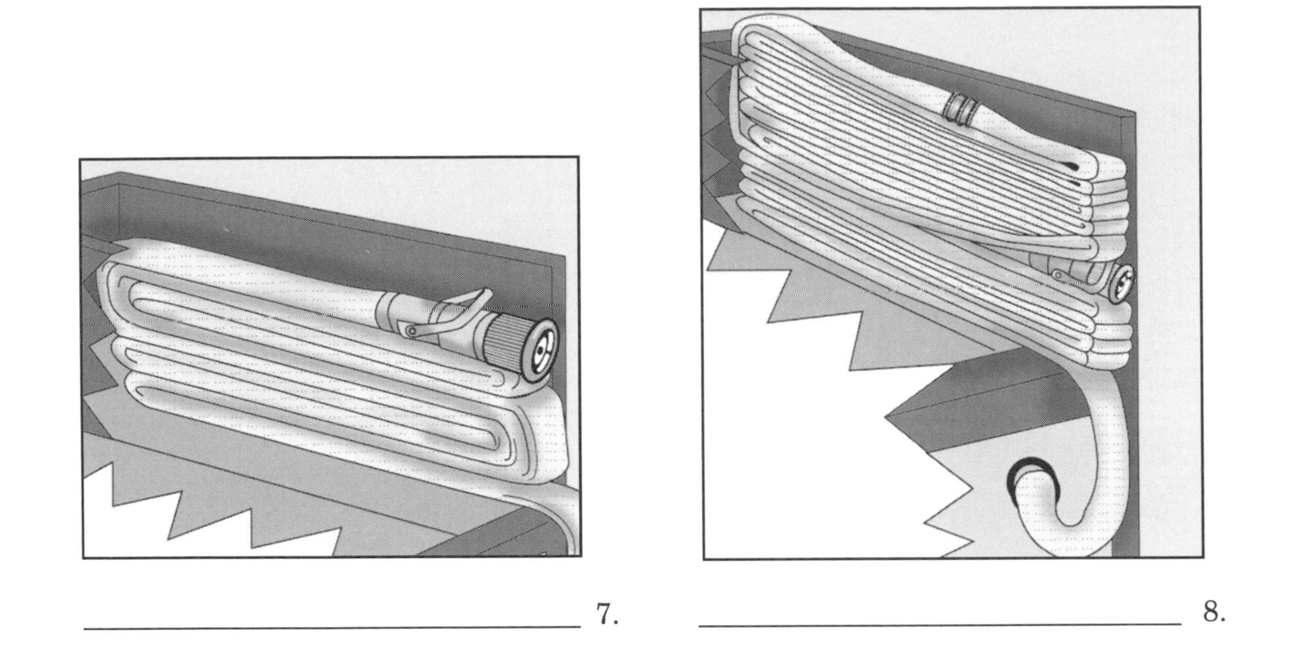

_______________ 7. _______________ 8.

G. Identify the following fire hose couplings. Write the correct name below each.

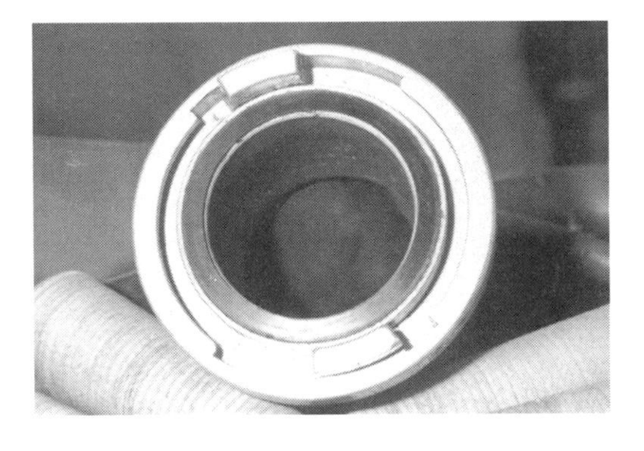

_______________ 1. _______________ 2.

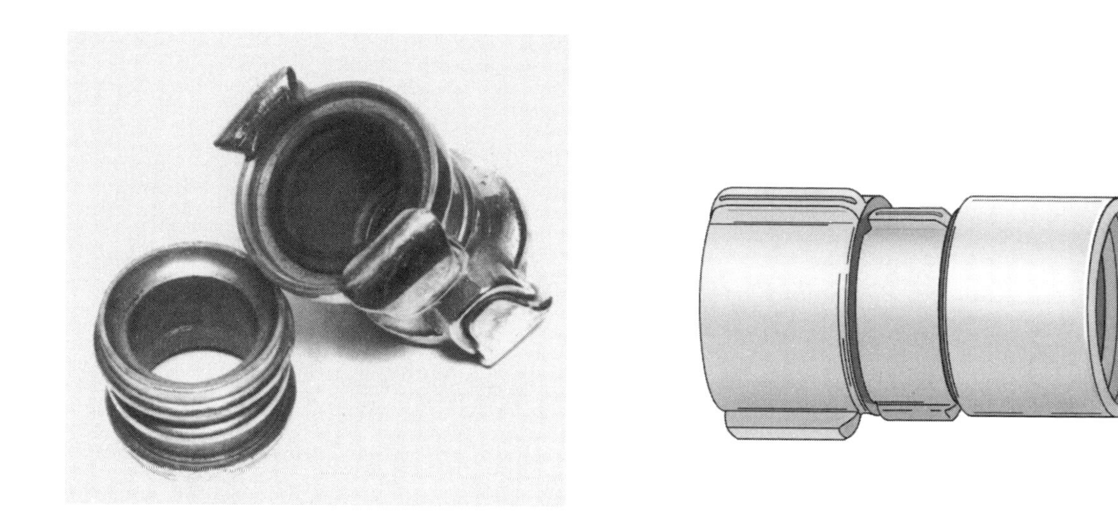

____________________________ 3.

____________________________ 4.

____________________________ 5.

FIREFIGHTER II

Matching

A. Match fire hose tools and appliances to their uses. Write the appropriate letters on the blanks.

____ 1. Connection that divides a line of hose into two or more lines

____ 2. Hardware accessories available for connecting hose of different sizes and thread types

____ 3. Intake device attached to the drafting end of a hard suction hose to keep debris from entering the fire pump

____ 4. Tool used for preventing hose from being damaged when dragged over sharp surfaces like roof edges and windowsills

____ 5. Two-piece metal cylinder used to repair a bad section on a hoseline that has ruptured

____ 6. Device used to tighten and loosen hose couplings

____ 7. Tool used primarily to remove caps from fire hydrant outlets and to open fire hydrant valves

____ 8. Tool used to strike lugs to tighten or loosen intake hose couplings

____ 9. Device used to help prevent injury to hose when vehicles cross it

____ 10. Device made to fit around fire hose where the hose is subjected to rubbing from vibrations

a. Fitting
b. Wye appliance
c. Hose roller
d. Hose bridge
e. Spanner
f. Chafing block
g. Suction hose strainer
h. Hydrant wrench
i. Hose jacket
j. Rubber mallet
k. Hose strap

B. **Match valves to their descriptions. Write the appropriate letters on the blanks.**

____ 1. Is open when the handle is in line with the hose; is closed when handle is at a right angle to the hose

____ 2. Has a baffle that is moved by a handle and screw arrangement

____ 3. Uses a flat baffle operated by a quarter-turn handle; baffle is in the center of the waterway when the valve is open

____ 4. Has hinged flat disk that swings in a door-like manner

a. Clapper
b. Ball
c. Wye
d. Gate
e. Butterfly

C. **Match fittings to their uses. Write the appropriate letters on the blanks.**

____ 1. Connects hose couplings with dissimilar threads but with the same inside diameter

____ 2. Extends a larger hoseline by connecting a smaller one to the end

____ 3. Changes the direction of water flow

____ 4. Closes off male couplings

____ 5. Closes off female couplings

a. Reducer
b. Splicer
c. Adapter
d. Hose plug
e. Hose cap
f. Elbow

True/False

D. **Write *True* or *False* before each of the following statements. Correct those statements that are false.**

________ 1. The difference between hose tools and hose appliances is that tools have water flowing through them and appliances do not.

____

____

________ 2. Clapper valves allow only one intake hose to be connected and charged before the addition of more hoses.

____

____

_____ 3. *Siamese appliance* and *wye appliance* are two names for the same device.

_____ 4. With increased use of large diameter hose, siamese appliances are now commonly being used to divide the large diameter hose into multiple smaller hoselines.

_____ 5. In general, large diameter hose appliances have one 4- or 5-inch *(100 mm or 125 mm)* inlet and two or more smaller outlets.

_____ 6. Even when using a hydrant valve, water flow of the original supply line must be interrupted to connect the supply pumper to the hydrant.

_____ 7. Double male or double female adapters are often used to connect hose when a pumper set up for a forward lay is used for a reverse lay.

_____ 8. Extending a hoseline with a reducer allows the option of adding another line if needed.

_____ 9. A hose jacket prevents hose from operating at full pressure.

_____ 10. Firefighters should stand directly over the handle of a hose clamp when applying or releasing it.

_____ 11. Pumper vibrations during fire fighting operations can keep intake hose in constant motion and cause wear.

____

____

_____ 12. Hose can be tested while it is bundled.

____

____

_____ 13. All personnel operating in a pressurized hose-testing area should wear safety helmets.

____

____

Multiple Choice

E. Write the letter of the best answer on the blank before each statement.

_____ 1. What wye connections are most common?

a. One 2-inch *(50 mm)* inlet to three 1¼-inch *(32 mm)* outlets
b. One 2½-inch *(65 mm)* inlet to two 1½-inch *(38 mm)* outlets
c. Two 2-inch *(50 mm)* inlets to three 1½-inch *(38 mm)* outlets
d. One 3-inch *(77 mm)* inlet to three 2-inch *(50 mm)* outlets

_____ 2. What does the typical siamese appliance feature?

a. Two or three female connections coming into the appliance and one male discharge exiting the appliance
b. One female connection coming into the appliance and two male discharges exiting the appliance
c. Two or three male connections coming into the appliance and one female discharge exiting the appliance
d. At least three male connections coming into the appliance and one or two female discharges exiting the appliance

_____ 3. How many incoming supply hoselines can be attached to a siamese appliance equipped with clapper valves?

a. One
b. Two
c. Three
d. Four

_____ 4. On what size hoseline is the water thief intended to be used?

a. 2-inch *(50 mm)* or smaller
b. 1½- to 3-inch *(38 mm to 77 mm)*
c. 2½-inch *(65 mm)* exclusively
d. 2½-inch *(65 mm)* or larger

____ 5. Which of the following devices is *not* a large diameter hose appliance?

a. Manifold
b. Water thief
c. Phantom pumper
d. Portable hydrant

____ 6. For what lays are hydrant valves used?

a. Forward
b. Reverse
c. Split
d. Combination

____ 7. In which of the following static water sources can a suction hose strainer rest on the bottom without damage?

a. Swift moving river
b. Shallow creek
c. Deep pond
d. Swimming pool

____ 8. In what two sizes are hose jackets made?

a. 1¾ and 2¾ inches *(45 mm and 70 mm)*
b. 2 and 2¾ inches *(50 mm and 70 mm)*
c. 2½ and 3 inches *(65 mm and 77 mm)*
d. 2¾ and 3¼ inches *(70 mm and 80 mm)*

____ 9. For what reason should a hose clamp be used to stop the flow of water in a hoseline?

a. Advancing a charged hoseline down a stairway
b. Attaching a charged hoseline to a standpipe
c. Advancing a charged hoseline up a ladder with only two firefighters
d. Replacing a burst section of hose without turning off the water supply

____ 10. What is a basic rule for attaching a hose clamp?

a. Apply the hose clamp no more than 15 feet *(4.6 m)* behind the apparatus.
b. Apply the hose clamp approximately 2 feet *(0.6 m)* from the coupling on the incoming water side.
c. Stand to one side when applying or releasing a hose clamp.
d. Open the clamp quickly to release excess pressure from the hose and avoid water hammer.

____ 11. Which fire hose tool often has other features, such as a prying wedge and a nail-pulling slot, already built into it?

a. Spanner wrench
b. Hydrant wrench
c. Rubber mallet
d. Hose bridge

____ 12. Who should perform acceptance testing on fire hose?

a. Safety officer
b. Manufacturer
c. Water supply officer
d. Anyone with Firefighter II certification

____ 13. How often must fire hose be performance tested?
- a. Monthly
- b. Biannually
- c. Quarterly
- d. Annually

____ 14. What should firefighters do with hose that is not repairable?
- a. Keep it on apparatus for backup hose.
- b. Take it out of service.
- c. Store it at the fire station for emergency use.
- d. Use it for training evolutions.

____ 15. What is a hose test gate valve?
- a. ¼-inch *(6 mm)* hole in the gate that permits pressurizing the hose but does not allow water to surge through the hose
- b. ½-inch *(13 mm)* opening in the gate that is used to bleed off hose pressure before beginning a fire attack
- c. ¾-inch *(19 mm)* slot in the gate that prevents a hoseline from rupturing
- d. 1-inch *(25 mm)* crevice in the gate that prevents a charged hoseline from whipping violently when the hose is loose and the nozzle is open

____ 16. How long should hose test lengths be?
- a. No less than 500 feet *(150 m)*
- b. No more than 200 feet *(60 m)*
- c. Between 300 feet and 500 feet *(90 m and 150 m)*
- d. No more than 300 feet *(90 m)*

Identify

F. Identify the following valves. Write the correct name below each.

____________________ 1.

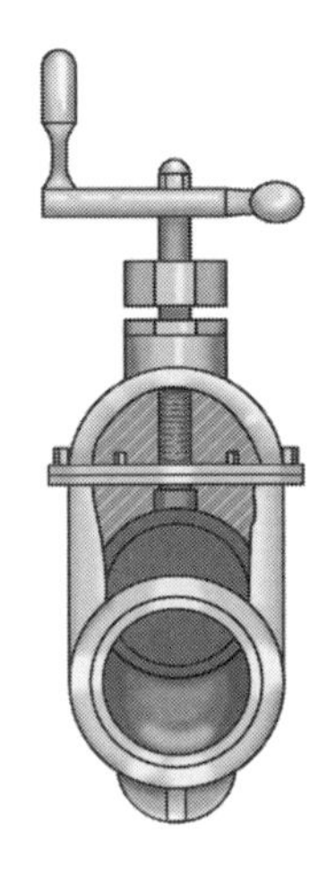

____________________ 2.

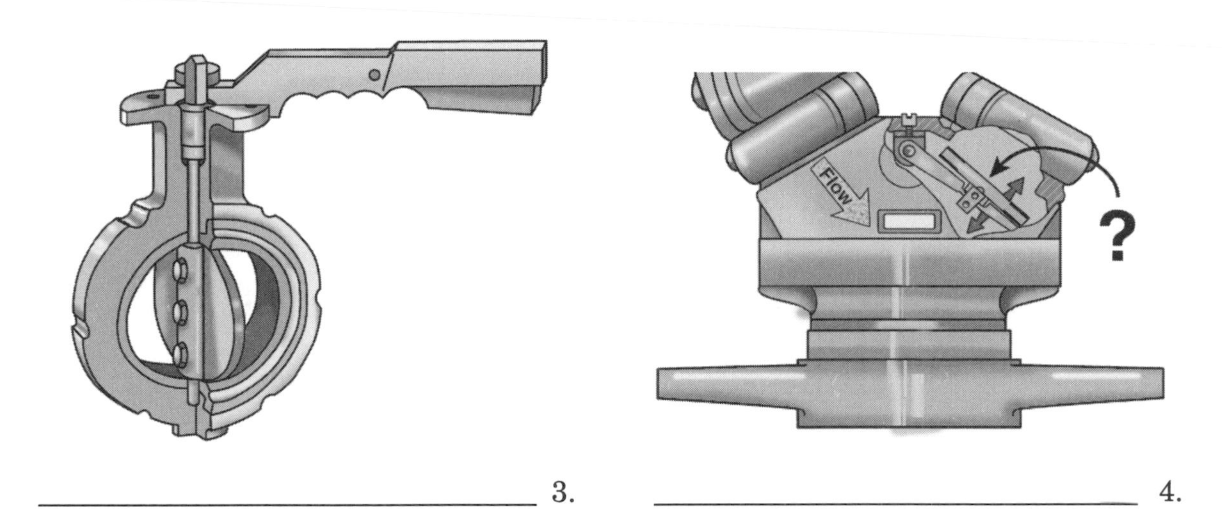

_______________ 3. _______________ 4.

G. **Identify the following appliances. Write the correct name below each.**

_______________ 1. _______________ 2.

_______________ 3. _______________ 4.

_________________________________ 5.

Chapter 13
Fire Streams

Chapter 13 Fire Streams

FIREFIGHTER I

Matching

A. Match to their definitions terms associated with fire streams. Write the appropriate letters on the blanks.

____ 1. Visible form of steam as it begins to cool

____ 2. Part of total pressure that is lost while forcing water through pipes, fittings, fire hose, and adapters

____ 3. Surge caused by the sudden stop of water flow through fire hose or pipe

____ 4. Volume of water flowing per minute

____ 5. Specific pattern of flowing water

____ 6. Fire stream produced from a fixed orifice, smoothbore nozzle

____ 7. Point at which water loses its forward velocity

____ 8. Force pushing back on a person handling a hoseline as water flows through a nozzle

____ 9. Fire stream composed of very fine water droplets

a. Solid stream
b. Friction loss
c. Breakover
d. Fire stream size
e. Fire stream type
f. Fog stream
g. Condensed steam
h. Nozzle reaction
i. Nozzle type
j. Water hammer

True/False

B. Write *True* or *False* before each of the following statements. Correct those statements that are false.

________ 1. Complete water vaporization occurs the instant that water reaches its boiling point.

____

____

_______ 2. Steam expansion is gradual.

____

____

_______ 3. Steam expansion can aid in reducing oxygen in a confined space thereby smothering a fire.

____

____

_______ 4. When a nozzle is above a fire pump, there is a pressure gain.

____

____

_______ 5. Water hammer cannot be avoided when working with fire hose.

____

____

_______ 6. Nozzles with flows in excess of 350 gpm *(1 400 L/min)* are not recommended for handlines.

____

____

_______ 7. Master streams are large-volume fire streams.

____

____

_______ 8. Solid streams should be used for energized electrical equipment fires.

____

____

_______ 9. A solid stream is also known as a *straight stream.*

____

____

_______ 10. Once the nozzle pressure has produced a stream with maximum reach, further nozzle pressure increases have little effect on the stream, except to increase the volume.

____

____

_______ 11. Fog streams are most useful for exterior fire fighting operations.

____

____

_______ 12. Firefighters must make necessary flow-rate adjustments *before* opening manually adjustable nozzles.

____

____

_______ 13. Major nozzle adjustments can throw a firefighter off balance.

____

____

_______ 14. Fire stream nozzles, in general, are not easy to control.

____

____

_______ 15. Broken streams are recommended for use on Class C fires.

____

____

_______ 16. The ball in a ball valve can be rotated a maximum of 45 degrees.

____

____

_______ 17. A ball valve will only operate when it is fully opened.

____

____

Multiple Choice

C. Write the letter of the best answer on the blank before each statement.

____ 1. At 212°F *(100°C)*, water expands approximately ____ times its original volume.
   a. 500
   b. 900
   c. 1,300
   d. 1,700

____ 2. What is true about water as a fire extinguishing agent?
   a. Water absorbs heat more quickly when more surface area is exposed.
   b. Water changes into steam with a relatively small amount of heat.
   c. Water has less heat-absorbing capacity than other common extinguishing agents.
   d. Water must be used simultaneously with foam to extinguish Class A fires.

____ 3. Which of the following guidelines helps reduce water pressure loss due to friction?
   a. Use long hoselines whenever possible.
   b. Use smaller hose or fewer lines when flow must be increased.
   c. Reduce flow by changing the nozzle tip or adjusting the flow setting.
   d. Keep nozzles and valves partially open when not operating hoselines.

____ 4. To what correlation does pumper elevation refer in fire fighting operations?
   a. The position of the static water source in relation to the pumper
   b. The position of the nozzle in relation to the pumping apparatus
   c. The height of the burning structure in relation to the pumper
   d. The altitude of the pumping apparatus in relation to sea level

____ 5. What sound does water hammer make?
   a. Loud roar, like rushing water
   b. Dull thud, like rubber mallet hitting solid wood
   c. Sharp clank, like a hammer striking a pipe
   d. High pitched whistle, like air escaping a balloon

____ 6. What fire stream is supplied by 1½- to 3-inch *(38 mm to 77 mm)* hose, which flows from 40 to 350 gpm *(160 L/min to 1 400 L/min)?*
   a. Handline stream
   b. Low-volume stream
   c. Mainline stream
   d. Master stream

____ 7. What fire stream discharges more than 350 gpm *(1 400 L/min)* and is fed by multiple 2½- or 3-inch *(65 mm or 77 mm)* hoselines?
   a. Handline stream
   b. Low-volume stream
   c. Mainline stream
   d. Master stream

____ 8. What fire stream discharges less than 40 gpm *(160 L/min)* including those fed by booster hoselines?

a. Handline stream
b. Low-volume stream
c. Mainline stream
d. Master stream

____ 9. At what psi *(kPa)* should solid stream nozzles be operated when used on handlines?

a. 35 *(245)*
b. 50 *(350)*
c. 70 *(490)*
d. 80 *(560)*

____ 10. At what psi *(kPa)* should solid stream nozzles be operated when used on a master stream device?

a. 35 *(245)*
b. 50 *(350)*
c. 70 *(490)*
d. 80 *(560)*

____ 11. Which of the following is *not* a characteristic of solid streams?

a. They provide more heat absorption per gallon *(liter)* delivered than other stream types.
b. They have greater reach than other stream types.
c. They are less likely to disturb normal layering of heat and gases during interior structural attacks than other stream types.
d. They do not allow for different stream pattern selections.

____ 12. What fire stream and nozzle pressure should be used on electrical equipment?

a. Solid stream, 50 psi *(350 kPa)*
b. Fog pattern, 70 psi *(490 kPa)*
c. Solid stream, 80 psi *(560 kPa)*
d. Fog pattern, 100 psi *(700 kPa)*

____ 13. What fire stream has the least forward velocity and shortest reach?

a. Solid
b. Straight fog
c. Wide-angle fog
d. Narrow-angle fog

____ 14. Which fire stream minimizes the chance of steam burns to firefighters?

a. Solid
b. Straight fog
c. Wide-angle fog
d. Narrow-angle fog

____ 15. What factor affects the reach of fog streams but *not* solid streams?

a. Gravity
b. Friction of air
c. Wind
d. Pattern selection

____ 16. How should firefighters adjust the rate of discharge from a manually adjustable fog nozzle?

a. Shut off the water flow and change nozzles.
b. Rotate the selector ring to a specific setting.
c. Open or close the nozzle partially.
d. Open or close the shutoff valve.

____ 17. To what flow rates can master streams be adjusted?
- a. 10 to 250 gpm *(40 L/min to 1 000 L/min)*
- b. 40 to 300 gpm *(160 L/min to 1 200 L/min)*
- c. 250 to 350 gpm *(1 000 L/min to 1 400 L/min)*
- d. 300 to 2,500 gpm *(1 200 L/min to 10 000 L/min)*

____ 18. For what purpose is the FLUSH setting on manually adjustable nozzles used?
- a. Opening the nozzle completely
- b. Allowing the largest flow rate
- c. Rinsing debris from the nozzle
- d. Draining the water from the hoseline

____ 19. What is true about a constant-pressure nozzle?
- a. The nozzleperson can change the flow rate by rotating the selector ring to a specific setting.
- b. The nozzle automatically varies flow rates to maintain effective nozzle pressure.
- c. The flow rate only changes if the water source pressure increases or decreases.
- d. The nozzleperson must keep fixed pressure on the nozzle to maintain the correct pressure.

____ 20. What is an advantage of using handline fog stream nozzles?
- a. Fog streams dissipate heat by exposing the maximum water surface for heat absorption.
- b. Fog streams are less susceptible to wind currents than are solid streams.
- c. Fog streams do not create heat inversion or cause steam burns to firefighters.
- d. Fog streams have more penetrating power than solid streams.

____ 21. What is true about broken streams?
- a. They absorb less heat per gallon *(liter)* than solid streams.
- b. They can conduct electricity.
- c. They have greater reach and penetration than fog streams.
- d. They should not be used on fires in confined spaces.

____ 22. What nozzle valve control operates by a screw that guides an exterior barrel forward or backward, rotating around an interior barrel?
- a. Ball
- b. Barrel
- c. Slide
- d. Rotary control

____ 23. Which nozzle valve is closed when the nozzle body is turned perpendicular to the waterway?
- a. Ball
- b. Barrel
- c. Slide
- d. Rotary control

____ 24. What valve seats a movable cylinder against a shaped cone to turn off the flow of water?

a. Ball
b. Barrel
c. Slide
d. Rotary control

____ 25. How should nozzles be cleaned?

a. Washed in a hose-washing machine while attached to hose
b. Washed with soap and water with a soft bristle brush
c. Soaked in vinegar and baking soda and rinsed with warm water
d. Scrubbed with steel wool and a mild detergent

Identify

D. Identify the following abbreviations associated with fire streams. Write the correct interpretation before each.

________________________________ 1. gpm

________________________________ 2. L/min

E. Identify the following nozzle control valves. Write the correct name below each.

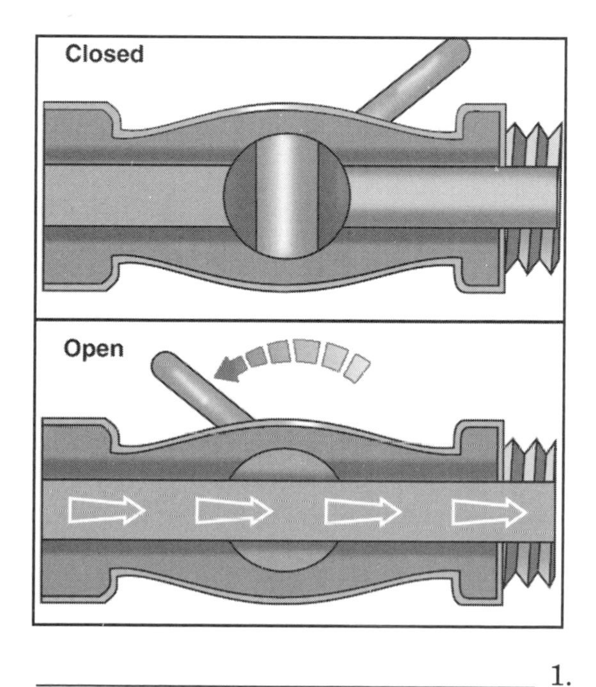

________________________ 1.

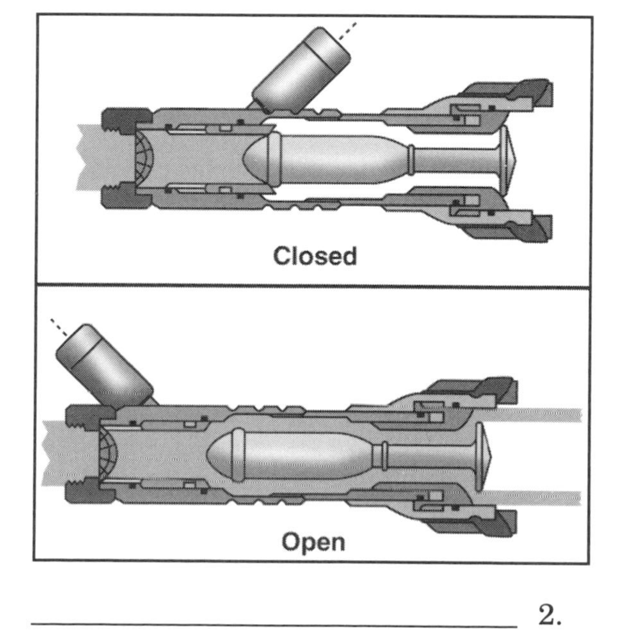

________________________ 2.

_______________________________ 3.

Photo for Question 3 courtesy of Brass Manufacturing Company.

FIREFIGHTER II

Matching

A. Match to their definitions terms associated with fire streams. Write the appropriate letters on the blanks.

____ 1. Raw foam liquid as it rests in its storage container before the introduction of water and air

____ 2. Device that introduces foam concentrate into the water stream to make the foam solution

____ 3. Mixture of foam concentrate and water before the introduction of air

____ 4. Completed product after air is introduced into the foam solution

____ 5. Foam mixture to which air has been added

a. Foam concentrate
b. Apportioned foam
c. Aerated foam
d. Foam solution
e. Foam proportioner
f. Finished foam

True/False

B. Write *True* or *False* before each of the following statements. Correct those statements that are false.

________ 1. Unignited chemicals have a tendency to either change the pH of water or remove water from fire fighting foams.

____

____

________ 2. Proper aeration produces foam bubbles of various sizes.

____

____

________ 3. Foam concentrates can be used interchangeably on burning fuels.

____

____

_______ 4. Increasing the surface tension of water in a foam solution makes the solution more effective.

____

____

_______ 5. Unignited spills do not require the same foam concentrate application rates as ignited spills.

____

____

_______ 6. Firefighters should periodically interrupt the foam application process to avoid applying too much solution.

____

____

_______ 7. Most foam concentrates can be mixed with salt water.

____

____

_______ 8. Firefighters should not adjust the proportioning percentage for Class A foams.

____

____

_______ 9. Firefighters may use any foam proportioner with any delivery device as long as both are made by the same manufacturer.

____

____

_______ 10. Injection method foam proportioning systems are commonly employed in apparatus-mounted or fixed fire protection system applications.

____

____

_______ 11. Batch-mixing is the simplest method of mixing foam concentrate and water.

____

____

_________ 12. Portable foam proportioners are generally inaccurate and seldom used in the fire service today.

____

____

_________ 13. Using a foam nozzle eductor compromises firefighter safety.

____

____

_________ 14. Firefighters should use fog nozzles with alcohol-resistant AFFF foams on polar solvent fires.

____

____

_________ 15. An air-aspirating foam nozzle has a greater stream reach than a standard fog nozzle.

____

____

_________ 16. Foam concentrates at full strengths pose significant health risks to firefighters.

____

____

_________ 17. Most Class A and Class B foam concentrates are mildly corrosive.

____

____

_________ 18. The more oxygen required to degrade a particular foam, the more environmentally friendly the foam is when it enters a body of water.

____

____

13

Multiple Choice

C. Write the letter of the best answer on the blank before each statement.

____ 1. What fire fighting foam action creates a barrier between fuel and fire?
   a. Shielding
   b. Suppressing
   c. Separating
   d. Cooling

____ 2. By what process does fire fighting foam prevent the release of flammable vapors and therefore reduce the possibility of ignition or reignition?
   a. Shielding
   b. Suppressing
   c. Separating
   d. Cooling

____ 3. For what two basic categories of flammable liquids are fire fighting foams especially effective?
   a. Hydrocarbon fuels and polar solvents
   b. Combustible liquids and self-reactive fuels
   c. Corrosive fuels and organic peroxides
   d. Compressed liquids and Class 9 fuels

____ 4. On which of the following liquids would fire fighting foam float?
   a. Alcohol
   b. Lacquer thinner
   c. Acetone
   d. Gasoline

____ 5. Which of the following liquids has an attraction for water, much like a positive magnetic pole attracts a negative pole?
   a. Crude oil
   b. Acid
   c. Benzene
   d. Naphtha

____ 6. What NFPA standard regulates low-expansion foam?
   a. 10
   b. 1971
   c. 11
   d. 1940

____ 7. What is the air/solution ratio for high-expansion foam?
   a. Up to 500 parts finished foam for every part of foam solution (500:1 ratio)
   b. At least 150 parts finished foam for every part of foam solution (150:1 ratio)
   c. From 75 to 200 parts finished foam for every part of foam solution (75:1 to 200:1)
   d. From 200 to 1,000 parts finished foam for every part of foam solution (200:1 to 1,000:1)

____ 8. Of what components is Class B foam made?

a. Hydrocarbon surfactants
b. Animal proteins
c. Organic peroxides
d. Oxygen asphyxiants

____ 9. Of what components is Class A foam made?

a. Hydrocarbon surfactants
b. Animal proteins
c. Organic peroxides
d. Oxygen asphyxiants

____ 10. What foam concentrate has supercleaning characteristics and is mildly corrosive?

a. Protein-based
b. Synthetic
c. Class A
d. Class B

____ 11. What effect would using a 6% foam at a 3% concentration have on fire suppression activities?

a. Adjusting the foam concentrate within 3 percentage points does not change the effectiveness of the foam.
b. Lowering the foam concentration decreases the amount of extinguishing agent needed to suppress a fire.
c. Reducing the concentration percentage decreases the harmful effects to the environment.
d. Trying to use foam in the wrong concentration will result in poor-quality foam.

____ 12. With what percentage of water are most foam concentrates intended to be mixed?

a. 0.1 to 6.0
b. 25.0 to 49.9
c. 67.0 to 82.5
d. 94.0 to 99.9

____ 13. What foam proportioning method uses the pressure energy in the stream of water to force foam concentrate into a fire stream?

a. Premixing
b. Injection
c. Induction
d. Batch-mixing

____ 14. What foam proportioning method uses an external pump or head pressure to force foam concentrate into a fire stream at the correct ratio in comparison to the flow?

a. Premixing
b. Injection
c. Induction
d. Batch-mixing

____ 15. What foam proportioning method mixes foam within a fire apparatus water tank or a portable water tank?

a. Premixing
b. Injection
c. Induction
d. Batch-mixing

____ 16. What foam proportioning method should be used only as a last resort with Class B foams?

a. Premixing
b. Injection
c. Induction
d. Batch-mixing

____ 17. What foam proportioning method is typically used with portable extinguishers, wheeled extinguishers, skid-mounted twin-agent units, and vehicle-mounted tank systems?

a. Premixing
b. Injection
c. Induction
d. Batch-mixing

____ 18. What foam proportioning method is limited to a one-time application?

a. Premixing
b. Injection
c. Induction
d. Batch-mixing

____ 19. What is the suction effect created as water at high pressure passes over a reduced opening and causes a low-pressure area near the outlet side of an eductor?

a. Vacuum effect
b. Venturi Principle
c. Nozzle reaction
d. Water pressure

____ 20. Where, in relation to the liquid surface of the foam concentrate, should the inlet to the in-line eductor be located?

a. No more than 3 feet *(1 m)* above
b. No more than 6 feet *(1.8 m)* above
c. Between 6 and 9 feet *(1.8 m and 2.7 m)* above
d. Exactly 9 feet *(2.7 m)* above

____ 21. What is true about foam nozzle eductors?

a. They are built into the nozzle.
b. They are built into the hoseline.
c. They can be attached to various nozzles.
d. They can be attached to any LDH.

____ 22. What device is defined as "*any nozzle that one to three firefighters can safely handle and that flows less than 350 gpm (1 400 L/min)*"?

a. Water-aspirating type nozzle
b. Mechanical blower generator
c. Handline nozzle
d. Medium-expansion device

____ 23. What nozzle type is limited to certain types of Class A applications?

a. Fog
b. Solid bore
c. Air-aspirating foam
d. Water-aspirating type

____ 24. Fixed-flow or automatic fog nozzles can be used with foam solutions to produce a ____.

a. Medium-expansion, long-lasting foam
b. Medium-expansion, short-lasting foam
c. Low-expansion, long-lasting foam
d. Low-expansion, short-lasting foam

____ 25. What foam generating device is typically associated with total-flooding applications?

a. Mechanical blower generator
b. Air-aspirating foam nozzle
c. Water-aspirating type nozzle
d. Solid bore nozzle

____ 26. What nozzle has a screen or series of screens that break up foam and mix it with air?

a. Fog
b. Solid bore
c. Air-aspirating foam
d. Water-aspirating type

____ 27. Which of the following is *not* a common reason for a foam fire stream system to fail to generate foam or to generate poor-quality foam?

a. Different foam concentration types are mixed in the same tank.
b. The hose lay on the discharge side of the eductor is too long.
c. The nozzle is not far enough above the eductor.
d. Air is leaking around fittings.

____ 28. What foam application method directs a foam stream on the ground near the front edge of a burning liquid pool?

a. Bank-down
b. Roll-on
c. Run-on
d. Rain-down

____ 29. What foam application method directs a foam stream into the air above the fire or spill and allows the foam to float gently down onto the surface?

a. Bank-down
b. Roll-on
c. Run-on
d. Rain-down

____ 30. What foam application method is directed off an elevated object, allowing foam to run down onto the surface of the fuel?

a. Bank-down
b. Roll-on
c. Run-on
d. Rain-down

____ 31. How is the biodegradability of a foam determined?

a. By the percentage of synthetic ingredients in it
b. By the rate at which oxygen chemically reacts with it
c. By the rate at which environmental bacteria cause it to decompose
d. By testing its effect on plant and marine life in lab settings

Identify

D. Identify the following abbreviations associated with fire streams. Write the correct interpretation before each.

_______________________________ 1. CAFS

_______________________________ 2. AFFF

_______________________________ 3. FFFP

_______________________________ 4. ARFF

_______________________________ 5. MSDS

E. Identify the following portable foam proportioners. Write the correct name below each.

_______________________________ 1.

_______________________________ 2.

F. Identify the following handline nozzles. Write the correct name below each.

_______________ 1.

_______________ 2.

Photo for Question 2 courtesy of Mount Shasta (CA) Fire Protection District.

Chapter 14
Fire Control

Chapter 14 Fire Control

FIREFIGHTER I

Matching

A. Match fire control terms to their definitions. Write the appropriate letters on the blanks.

____ 1. Master stream device movement caused by pressure in hoselines

____ 2. Cancer-causing substance

____ 3. Electrical current in excess of 600 volts

____ 4. Tendency of an energized electrical conductor to pass its current along the path of least resistance

a. Electrical exposure
b. Carcinogen
c. Ground gradient
d. Crawling away
e. High voltage

B. Match parts of a wildland fire to their definitions. Write the appropriate letters on the blanks. Definitions are continued on the next page.

____ 1. The area where the fire started; the point from which fire spreads

____ 2. The part of a wildland fire that travels or spreads most rapidly

____ 3. A long, narrow strip of fire extending from the main fire

____ 4. The outer boundary, or the distance around the outside edge, of the burning area

____ 5. The side opposite the head; usually burns slowly and quietly

____ 6. The side roughly parallel to the main direction of fire spread

____ 7. An unburned area inside the fire perimeter

a. Heel
b. Spot fire
c. Head
d. Black
e. Flank
f. Origin
g. Island
h. Perimeter
i. Finger
j. Green

____ 8. The area of unburned fuels next to the involved area

____ 9. The area in which the fire has consumed the fuels

C. Match wildland fire terms to their definitions. Write the appropriate letters on the blanks.

____ 1. The compass direction a slope faces

____ 2. A terrain feature that creates turbulent updrafts causing a chimney effect

____ 3. A steep V-shaped drainage

____ 4. A depression between two adjacent hilltops

____ 5. The fire attack made against the flames at the edge of a wildland fire

____ 6. The fire attack used at varying distances from an advancing fire to halt its progress

a. Aspect
b. Rock outcropping
c. Indirect attack
d. Direct attack
e. Saddle
f. Drainage
g. Chute

True/False

D. Write *True* or *False* before each of the following statements. Correct those statements that are false.

________ 1. Regardless of the conditions at a fire scene, firefighters should immediately begin a fire attack.

____

____

________ 2. Firefighters should locate the main fire area before stopping to extinguish burning fascia and soffit, boxed cornices, or other doorway overhangings.

____

____

_____ 3. During a direct attack in a structure, firefighters should apply water in short bursts directly on the burning fuels until the fire "darkens down."

_____

_____

_____ 4. Firefighters should apply water continuously for an extended period of time during a direct attack in a structure to avoid upsetting thermal layering.

_____

_____

_____ 5. Firefighters should choose an indirect fire attack for areas where victims may still be trapped.

_____

_____

_____ 6. Master stream devices operate at high flow rates, which in turn means large amounts of friction loss.

_____

_____

_____ 7. The primary danger of electrical fires is the failure of firefighters to disconnect the energy source.

_____

_____

_____ 8. Commercial buildings with electrically operated elevators and/or air handling equipment should not be unilaterally de-energized.

_____

_____

_____ 9. Firefighters should stop the flow of electricity to any object involved in fire before initiating fire-suppression activities.

_____

_____

_____ 10. Transformer fires create few health or environmental risks.

_____

_____

_________ 11. Firefighters should pull the electrical meter to turn off the electricity in residential fires.

____

____

_________ 12. Firefighters should not look directly at an arcing electrical line.

____

____

_________ 13. Electrical control boxes should be left open to indicate that they are out of service.

____

____

_________ 14. Downed electrical lines are dangerous only when they are arcing.

____

____

_________ 15. If a difference in electrical potential exists between a firefighter's feet and an object dragged along the ground, a current may pass through the firefighter and return to the ground through the dragged object.

____

____

_________ 16. A combustible metal fire is extinguished when flames are no longer visible.

____

____

_________ 17. If smoke or fire is visible as the first-due company approaches the scene, firefighters should stop and lay a supply line from a hydrant to the scene.

____

____

_________ 18. The truck company should initially begin laying hose for the fire attack.

____

____

_________ 19. When fighting a vehicle fire, firefighters should immediately open the vehicle's hood to effectively knock down an engine compartment fire.

_________ 20. Passenger vehicles do not contain extraordinary hazards.

_________ 21. Medium- and large-sized wildland fires may create their own winds.

_________ 22. Precipitation largely determines the moisture content of live fuels.

_________ 23. Wildland fires usually move faster downhill than uphill.

_________ 24. At a wildland fire, "green" refers to a safe area.

_________ 25. Firefighters must perform size-up periodically during a wildland fire.

_________ 26. In every case where a firefighter was killed while fighting a wildland fire, one or more of the Ten Standard Orders had been ignored.

Multiple Choice

E. Write the letter of the best answer on the blank before each statement.

____ 1. What tools should hose advancing teams carry?
   a. Portable fan, claw hammer, and axe
   b. Portable light, prying tool, and axe
   c. Pneumatic drill, maul, and axe
   d. Power saw, mallet, and axe

____ 2. What should be done with a burning mattress during a fire attack?
   a. Take it outside the building for extinguishment.
   b. Move it into a hallway for extinguishment.
   c. Extinguish it where it is found.
   d. Leave it alone to burn itself out.

____ 3. What fire stream should be selected when firefighters can make adequate ventilation openings ahead of the nozzles?
   a. Narrow fog pattern
   b. Wide fog pattern
   c. Straight stream
   d. Broken stream

____ 4. How does discharging water at smoke affect fire fighting operations?
   a. Increases visibility and decreases thermal-layering disruption
   b. Increases steam production and decreases fire spread
   c. Increases visibility and decreases fire spread
   d. Increases water damage and decreases visibility

____ 5. Where should a fire stream be directed when a fire is localized?
   a. Above the fire
   b. Into the middle of the fire
   c. At the base of the fire
   d. Around the edges of the fire

____ 6. For which of the following fires would a booster line be inappropriate?
   a. Well-involved structure fire
   b. Small brush fire
   c. Small exterior fire
   d. Dumpster® trash container fire

____ 7. For which of the following situations would a 1½ inch *(38 mm)* hoseline be used?
   a. When large volume and great reach are required for exposure protection
   b. When maximum reach required does not exceed 50 feet *(15 m)*
   c. When two or more floors are fully involved
   d. When 125–350 gpm *(500 L/min to 1 400 L/min)* water is needed

____ 8. During an indirect attack, what fire stream should *not* be played back and forth in the superheated gases at the ceiling level?

a. Solid
b. Straight
c. Narrow fog
d. Wide fog

____ 9. What nozzle pattern should be used during a combination method structure fire attack?

a. C
b. M
c. Z
d. H

____ 10. What common nozzle pattern is achieved by rotating the stream edge to reach the ceiling, wall, floor, and opposite wall?

a. O
b. D
c. Q
d. U

____ 11. Which of the following statements is true about moving a master stream device once the line is in operation?

a. It must be clamped and loaded on an apparatus before it is moved.
b. It must be shut down to be moved.
c. It takes a team of four firefighters to carry an operating line.
d. It must be strapped to firefighters' legs with leg straps.

____ 12. At what angle should a master stream device enter a burning structure?

a. At a downward angle and deflect off the floor
b. At a diagonal angle and deflect off the walls
c. At an upward angle and deflect off the ceiling
d. Parallel to the floor without deflecting off any object

____ 13. What is the smallest hoseline that should be used with a master stream appliance?

a. 1½-inch *(38 mm)*
b. 1¾-inch *(45 mm)*
c. 2¼-inch *(60 mm)*
d. 2½-inch *(65 mm)*

____ 14. What is the minimum number of firefighters needed to deploy a master stream device and supply water to it?

a. One
b. Two
c. Three
d. Four

____ 15. In which of the following situations may a master stream device be anchored and unmanned?

a. Never
b. During a fire in a building with more than two stories
c. During a fabric warehouse fire
d. At an LPG storage tank fire

____ 16. Whose responsibility is it to ensure that appropriate power breakers are opened to control power flow to involved structures?
  a. Any firefighter with Firefighter I training
  b. Safety officer
  c. Fire officer
  d. Incident commander

____ 17. What extinguishing agent is best for fires in sensitive electronic or computer equipment?
  a. Halon
  b. Dry chemical
  c. Water
  d. Dry powder

____ 18. What size area should be cleared on either side of a broken electrical transmission line?
  a. Area equal to half of the span between poles
  b. Area equal to the span between poles
  c. Area approximately 10 × 10 feet *(3 m by 3 m)*
  d. Area no less than 15 × 15 feet *(5 m by 5 m)*

____ 19. Who should cut live electrical wires on a fireground?
  a. Any firefighter with Firefighter I training
  b. Fire officer
  c. Power company personnel
  d. Any person with appropriate equipment

____ 20. How should aboveground transformers be extinguished?
  a. From the ground with a master stream device
  b. From the ground with a wheeled carbon dioxide extinguisher
  c. From an aerial device with a halon extinguisher
  d. From an aerial device with a dry chemical extinguisher

____ 21. What is the most frequent hazard associated with underground transmission systems?
  a. Implosion
  b. Explosion
  c. Electrocution
  d. Power outage

____ 22. Which of the following actions should be taken during a utility vault fire?
  a. Discharge dry chemical agent into the vault and leave the cover off.
  b. Extinguish the fire with a solid water stream.
  c. Enter the vault and apply dry powder directly on the fire.
  d. Discharge carbon dioxide into the vault and replace the cover.

____ 23. Which of the following actions should *not* be taken at a high-voltage installation fire?

a. Use a water fog stream to extinguish the fire.
b. Wear self-contained breathing apparatus.
c. Use a safety line monitored by someone outside the enclosure.
d. Perform search and rescue operations with a clinched fist or the back of the hand.

____ 24. When should appropriate personnel shut down electrical power during structural fire fighting operations?

a. Immediately upon arrival
b. After completing rescue operations
c. When an electrical hazard exists
d. Before ventilation procedures are performed

____ 25. What does a lockout device do?

a. Ensures that electrical power will not inadvertently restore after being turned off
b. Enables firefighters to turn the electrical power on and off as needed
c. Prevents people from tampering with and turning off electrical power
d. Indicates that the electrical equipment is energized

____ 26. How should a firefighter dismount from an apparatus that may be electrically charged?

a. Use a wooden ladder to climb down.
b. Jump clear of the apparatus.
c. Climb off the back of the apparatus.
d. Hold the handrails and step off one foot at a time.

____ 27. Which of the following statements is true about Class D fires?

a. Water is most effective applied in small amounts and intermittently.
b. Combustible metal fires can contain temperatures greater than 2,000°F *(1 093°C)* even if they appear suppressed.
c. Directing hose streams at burning metal knocks down the fire most efficiently.
d. Combustible metal fires can be recognized by a characteristic orange glow that is given off.

____ 28. What engine company usually initiates incident command and the fire attack?

a. First-due engine company
b. Second-due engine company
c. Truck company
d. Rescue company

____ 29. Which of the following responsibilities usually belongs to the truck company?
a. Back up the initial attack lines.
b. Protect the secondary means of egress.
c. Prevent fire extension.
d. Set up lighting equipment.

____ 30. Which of the following statements is appropriate to vehicle fires?
a. The attack line should be at least a 2-inch *(50 mm)* hoseline.
b. Booster lines provide adequate protection and rapid cooling.
c. Portable extinguishers may be used for carburetor fires.
d. The fire should be attacked from the downhill side.

____ 31. What is the external temperature of a properly operating catalytic converter?
a. 500°F *(260°C)*
b. 800°F *(427°C)*
c. 1,300°F *(704°C)*
d. 2,500°F *(1 371°C)*

____ 32. What are the three most important factors once a wildland fire starts?
a. Weather, personnel, and daylight
b. Fuel, weather, and topography
c. Personnel, extinguishing agent, and weather
d. Topography, extinguishing agent, and fuel

____ 33. How does the compactness of fuel affect a wildland fire?
a. It makes the fuel burn faster.
b. It makes the fuel burn slower.
c. It influences the amount of water needed to perform extinguishment.
d. It increases the temperature at which the fuel will ignite.

____ 34. Which of the following weather aspects primarily affects wildland fuels as a result of long-term drying?
a. Temperature
b. Relative humidity
c. Wind
d. Precipitation

Identify

F. Identify the following abbreviations. Write the correct interpretation before each.

____________________ 1. LPG

____________________ 2. PCB

____ 3. SRS

____ 4. SIPS

List

G. List the Ten Standard Fire Fighting Orders. Write the correct fire order after each appropriate first letter.

1. **F** ____
2. **I** ____
3. **R** ____
4. **E** ____
5. **O** ____
6. **R** ____
7. **D** ____
8. **E** ____
9. **R** ____
10. **S** ____

FIREFIGHTER II

Matching

A. Match fire control terms to their definitions. Write the appropriate letters on the blanks.

____ 1. Fire that involves flammable and combustible liquids and gases

____ 2. Liquid with a flash point less than 100°F *(38°C)*

____ 3. Liquid with a flash point higher than 100°F *(38°C)*

____ 4. Rupture in a tank caused by the sudden release and consequent vaporization of liquids

____ 5. Fuel gas stored in a liquid state under pressure

a. Combustible liquid
b. Flammable liquid
c. LPG
d. BLEVE
e. Class B fire
f. Class D fire

True/False

B. Write *True* or *False* before each of the following statements. Correct those statements that are false.

________ 1. Firefighters should avoid standing in a pool of runoff water while fighting a fire involving flammable or combustible liquids.

____

____

________ 2. Firefighters should extinguish fires burning around a relief valve or piping if the leaking product cannot be turned off.

____

____

________ 3. Firefighters should depend on relief valves as a safe way to relieve excess tank pressure during severe fire conditions.

____

____

_______ 4. Experience has shown that water is ineffective as an extinguishing agent for Class B fires.

____

____

_______ 5. Firefighters should not flush Class B fuels down drains or sewers.

____

____

_______ 6. Water can be used to displace fuel from pipes or tanks that are leaking.

____

____

_______ 7. Firefighters should learn to use hoselines as protective covers during on-the-job training.

____

____

_______ 8. When a liquid fuel tank is exposed to fire, firefighters should approach the tank from the end.

____

____

_______ 9. Compressed natural gas is not subject to BLEVE.

____

____

_______ 10. When a break in a gas utility occurs, firefighters should contact the utility company immediately even if the gas has already ignited.

____

____

_______ 11. Firefighters should adjust gas main valves to shut off fuel to broken or damaged lines.

____

____

_________ 12. The rapid intervention crew may be assigned to emergency scene duties other than personnel rescue.

_________ 13. Upon arriving at the scene, a chief officer may choose to assume command from the first arriving incident commander.

_________ 14. The heat conditions of a basement fire are much more intense than those of a standard structural fire.

_________ 15. Even with proper ventilation, firefighters must be aware of the dangers of poor visibility during any belowground fire.

_________ 16. Immediately upon arrival at the scene of a fire in an enclosed or confined area, firefighters should enter the area and begin a fire attack.

Multiple Choice

C. Write the letter of the best answer on the blank before each statement.

____ 1. Where should water be applied to a tank that is in danger of rupturing?
- a. To the base
- b. Around the middle section
- c. On the bottom half
- d. To the upper portions

____ 2. What is the preferred extinguishing agent to control flammable liquid fires?
- a. Water
- b. Foam
- c. Halon
- d. Dry powder

____ 3. For which of the following Class B fuels can water in droplet form be used to extinguish fire?

a. Raw crude
b. Gasoline
c. Kerosene
d. Alcohol

____ 4. What fire stream can be applied directly to a small opening on a fuel tank to keep escaping liquid back?

a. Solid
b. Broken
c. Fog
d. Straight

____ 5. What fire stream can be used to dissipate flammable vapors?

a. Solid
b. Broken
c. Fog
d. Straight

____ 6. What factor is *not* involved in the placement of apparatus at a bulk transport vehicle emergency?

a. Topography
b. Weather conditions
c. Traffic
d. Speed limit

____ 7. How can firefighters determine the exact nature of the cargo carried by a truck involved in a leak or fire?

a. Looking at the substance
b. Smelling the substance
c. Testing the substance
d. Reading the placard

____ 8. Which of the following statements describes natural gas?

a. It is made of mostly butane with small quantities of ethane and methane added.
b. It has no odor of its own.
c. It is heavier than air.
d. It is classified as an asphyxiant because it is a toxic substance.

____ 9. At what concentration is natural gas explosive?

a. 3 percent
b. 4 percent
c. Between 2 and 5 percent
d. Between 5 and 15 percent

____ 10. Which of the following statements describes bottled gas?

a. It is stored in a liquid state under pressure.
b. It is composed of mainly methane with small quantities of propane and butane added.
c. It is lighter than air.
d. It has a pungent odor of its own.

____ 11. At what concentration is bottled gas explosive?

a. 0.5 percent
b. 0.75 percent
c. Between 1 and 1.5 percent
d. Between 1.5 and 10 percent

____ 12. What fire stream type and size is necessary to dissipate unburned gas?
   a. Solid stream of 50 gpm *(200 L/min)*
   b. Fog stream of 75 gpm *(300 L/min)*
   c. Solid stream of 100 gpm *(400 L/min)*
   d. Fog stream of 100 gpm *(400 L/min)*

____ 13. What action should a firefighter take when faced with a broken underground gas line?
   a. Evacuate the upwind areas first.
   b. Turn off the operating valves to the gas line.
   c. Protect exposures with hose streams.
   d. Extinguish the burning gas.

____ 14. Who determines the exact number of rapid intervention crews needed at an emergency scene?
   a. Fire officer
   b. Safety officer
   c. Incident commander
   d. Any firefighter with at least Firefighter II training

____ 15. At what temperature do unprotected steel supports elongate?
   a. 575°F *(302°C)*
   b. 650°F *(343°C)*
   c. 825°F *(441°C)*
   d. 1,000°F *(538°C)*

____ 16. Which of the following tools should *not* be used to indirectly attack belowground fires?
   a. Booster line
   b. Piercing nozzle
   c. Cellar nozzle
   d. Distributor

____ 17. What can be used to extinguish a fire in a basement that is inaccessible?
   a. Halon
   b. Dry chemicals
   c. High-expansion foam
   d. Water

____ 18. What precaution should be taken when firefighters enter a confined enclosure to extinguish a fire?
   a. A lifeline should be tied to each pair of firefighters.
   b. Teams on the inside and outside should depend on portable radios for communication.
   c. A standby crew equal to at least one-half the original crew should be available.
   d. Crews should determine a backup communication system.

____ 19. Which of the following signals is accurate according to the *O-A-T-H* method of communication?

a. One tug means *advance.*

b. Three tugs means *take up.*

c. *A* means *assist.*

d. *H* means *hurry.*

____ 20. What is the best way to monitor the atmosphere of a confined space for rescuer and firefighter safety?

a. Use atmosphere monitoring devices.

b. Use combustible-gas meter readings.

c. Disregard monitoring devices and always wear SCBA.

d. Send a parakeet into the area.

____ 21. Which of the following items does *not* have to be recorded by the safety officer as a firefighter or rescuer enters a hazardous area?

a. Name

b. Estimated safe working time

c. Certification level

d. Tank pressure

Identify

D. Identify the following abbreviations associated with fire control. Write the correct interpretation before each.

________________________ 1. BLEVE

________________________ 2. LPG

________________________ 3. CNG

________________________ 4. RIC

________________________ 5. ppm

Chapter 15
Fire Detection, Alarm, and Suppression Systems

Chapter 15 Fire Detection, Alarm, and Suppression Systems

FIREFIGHTER I

True/False

A. **Write *True* or *False* before each of the following statements. Correct those statements that are false.**

_____ 1. Pumpers should not connect to the fire department connection unless the automatic sprinkler system fails to extinguish the fire.

____

____

_____ 2. When a sprinkler valve is closed, a firefighter with a portable radio should be stationed at the valve in case it needs to be reopened.

____

____

_____ 3. Fire department personnel service residential and commercial automatic sprinkler systems.

____

____

_____ 4. To stop the flow of water from a sprinkler, firefighters can insert a wedge between the discharge orifice and the deflector, and then tap the wedge by hand until the flow stops.

____

____

_____ 5. Sprinkler system control valves should remain closed until firefighters connect hoselines to the fire department connection.

____

____

Identify

B. **Identify the main control valve in the sprinkler system illustrated below. Circle the letter next to the arrow pointing to the correct valve.**

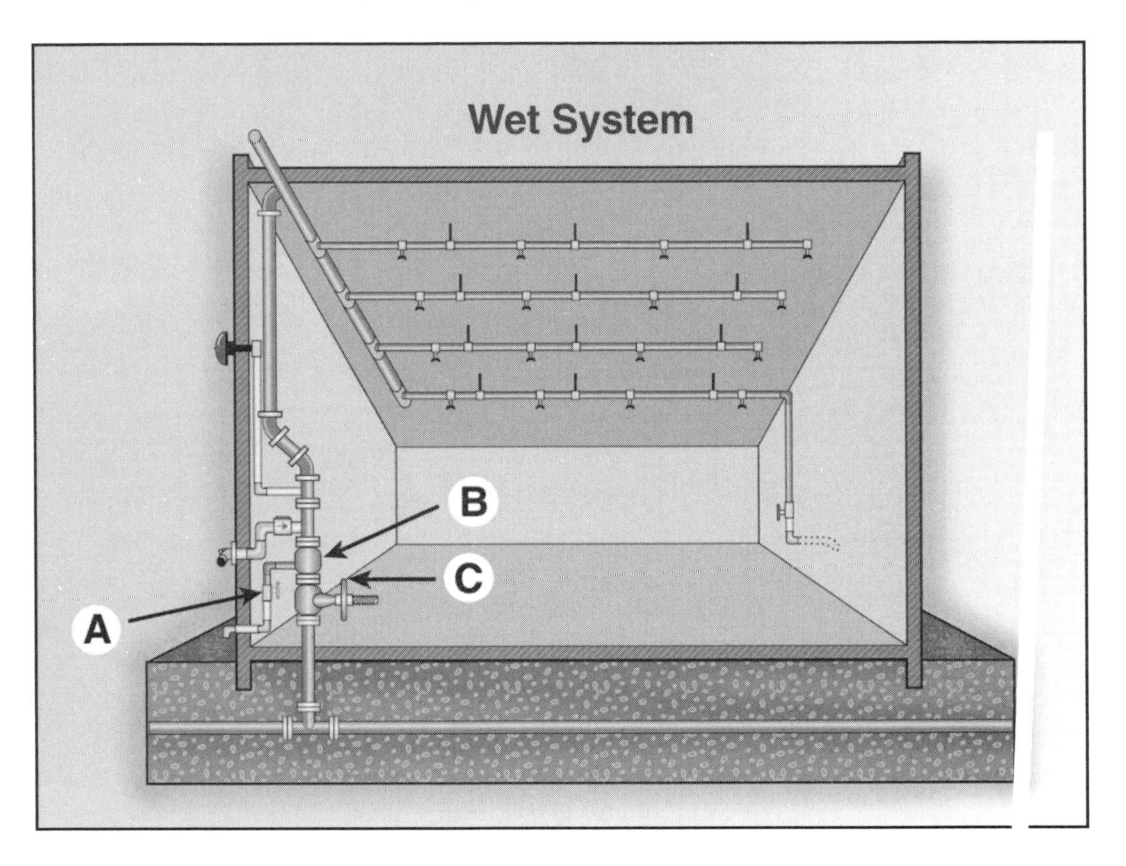

FIREFIGHTER II

Matching

A. Match alarm systems and detection devices to their definitions. Write the appropriate letters on the blanks.

____ 1. Alarm system designed to be initiated only manually

____ 2. Detection device that holds a spring-operated contact open with a solder that has a known melting temperature

____ 3. Device that holds electrical contacts apart and is designed to break when the liquid inside is heated to a predetermined temperature

____ 4. Device designed to detect heat only in a relatively small area surrounding the specific spot in which it is located

____ 5. Device that detects heat over a linear area parallel to the detector

____ 6. Device designed to initiate a signal when the rise of temperature in a room exceeds normal amounts

____ 7. Device designed for use in areas that are normally subject to regular temperature changes that are slower than those under fire conditions

____ 8. Auxiliary alarm system within an occupancy that is attached directly to a hard-wired or radio-type municipal fire alarm box

____ 9. Alarm system in which the municipal alarm circuit extends into the protected property

a. Spot-type detector

b. Fusible device

c. Local energy system

d. Rate-of-rise heat detector

e. Local system

f. Remote station system

g. Rate-compensated detector

h. Continuous line detection device

i. Shunt system

j. Frangible-bulb device

B. **Match automatic sprinkler system terms to their definitions. Write the appropriate letters on the blanks.**

____ 1. The vertical piping to which a sprinkler alarm valve, control valve, and other components are attached

____ 2. The pipe connecting the riser to the cross mains

____ 3. The pipe that directly feeds the lines to which sprinklers are installed

____ 4. A fixed-spray nozzle that is operated individually by a thermal detector

____ 5. The device that catches excess water from momentary water pressure surges

____ 6. The device that keeps water out of the sprinkler piping until a fire actuates a sprinkler

a. Cross main
b. Sprinkler
c. Retard chamber
d. Feed main
e. Branch line
f. Riser
g. Dry-pipe valve

C. **Match sprinkler system types to their definitions. Write the appropriate letters on the blanks.**

____ 1. System used in locations that will not be subjected to temperatures below 40°F *(4°C)*

____ 2. System that has a backflow prevention check valve and an electronic flow alarm

____ 3. System used in locations where piping may be subjected to temperatures below 40°F *(4°C)*

____ 4. Dry system that employs a deluge-type valve

____ 5. System that wets down an area where a fire originates by discharging water from all open heads in the system

a. Dry-pipe
b. Straight stick
c. Deluge
d. Wet-pipe
e. Preaction
f. Overflow

True/False

D. Write *True* or *False* before each of the following statements. Correct those statements that are false.

_______ 1. Most state-of-the-art fire detection and alarm systems operate electronically.

____

____

_______ 2. Frangible-bulb detectors are the most common detectors manufactured today.

____

____

_______ 3. Bimetallic detectors must be replaced to restore a system after use.

____

____

_______ 4. Heat detectors are preferred for most occupancies because they detect fire much more quickly than smoke detectors.

____

____

_______ 5. Photoelectric smoke detectors are generally more sensitive to smoldering fires than are ionization smoke detectors.

____

____

_______ 6. The increase in current flowing between the plates of an ionization smoke detector initiates an alarm signal.

____

____

_______ 7. Fire detectors powered by household current are the most reliable mechanisms for rural areas where power failures are frequent.

____

____

_________ 8. The alarm signal in a shunt system is instantly transmitted to the alarm center over the municipal system.

____

____

_________ 9. Fire alarm systems are designed to be self-supervising.

____

____

_________ 10. Pneumatically activated carbon dioxide systems should be removed from service.

____

____

_________ 11. Automatic sprinklers and all component parts must be listed by a nationally recognized testing laboratory.

____

____

_________ 12. Sprinklers designed for different positions are interchangeable.

____

____

_________ 13. Control valves in sprinkler systems are located at the source of water supply.

____

____

_________ 14. The main control valve should always be returned to the closed position after maintenance is complete.

____

____

_________ 15. The proper direction of water flow through a check valve may be denoted by arrows on the valve or by observing the appearance of the valve casting.

____

____

_______ 16. A deluge system does not need a supplemental detection system because the sprinklers have heat-responsive elements in them.

____

____

Multiple Choice

E. Write the letter of the best answer on the blank before each statement.

____ 1. Where should heat detectors be placed?

a. On the floor
b. Above the baseboards on a wall
c. At eye level on a wall
d. On the ceiling

____ 2. Which of the following detection devices restores itself when the level of heat is reduced?

a. Frangible-bulb
b. Tubing-type
c. Fusible device
d. Influx-type

____ 3. A rate-of-rise heat detector initiates a signal when the rise in temperature exceeds ____.

a. 9 to 12°F *(6°C to 7°C)* above normal room temperature
b. 12 to 15°F *(7°C to 8°C)* above normal room temperature
c. 9 to 12°F *(6°C to 7°C)* per minute
d. 12 to 15°F *(7°C to 8°C)* per minute

____ 4. What is the most common type of rate-of-rise detector?

a. Pneumatic spot detector
b. Electric line detector
c. Battery-operated bimetallic detector
d. Photoelectric detector

____ 5. What is the maximum recommended length of tubing in a line detector system?

a. 500 feet *(150 m)*
b. 800 feet *(240 m)*
c. 925 feet *(280 m)*
d. 1,000 feet *(300 m)*

____ 6. How should the tubing in a line detector system be arranged?

a. In rows that are not more than 45 feet *(14 m)* long
b. In rows that are at least 30 feet *(9 m)* long
c. In rows that are not more than 15 feet *(5 m)* from walls
d. In rows that are at least 30 feet *(9 m)* from walls

FIREFIGHTER II

____ 7. Which of the following applications is one way that a photoelectric cell detects smoke?
   a. Beam
   b. Radiation
   c. Conduction
   d. Convection

____ 8. How often should batteries in a smoke detector be changed?
   a. No more than once per year
   b. No less than twice per year
   c. Approximately once every two years
   d. As needed or every two years

____ 9. Which of the following light sources would *not* activate a single-band IR detector?
   a. Sunlight
   b. Welding arcs
   c. Halogen bulbs
   d. Mercury-vapor lamps

____ 10. Which of the following gases is most practical to monitor for fire-detection purposes?
   a. Carbon dioxide
   b. Hydrogen chloride
   c. Hydrogen cyanide
   d. Hydrogen sulfide

____ 11. In what type of community would a local energy system be used?
   a. One served by a remote station system
   b. One served by a parallel telephone system
   c. One served by a shunt system
   d. One served by a municipal fire-alarm-box system

____ 12. What NFPA standard requires a predischarge alarm for carbon dioxide extinguishing systems?
   a. 14A
   b. 13D
   c. 12
   d. 10

____ 13. What information should be included on the pre-incident survey with regard to the sprinkler system?
   a. Age of building
   b. Type of occupancy
   c. Size of building
   d. Fire extension possibilities

____ 14. What NFPA standard addresses the primary guidelines for installation of sprinkler protection in occupancies?
   a. 13D
   b. 1910
   c. 14A
   d. 1970

_____ 15. For what reason is a sprinkler system *least* likely to fail?
a. Interruption to the municipal water supply
b. Failure of the actual sprinklers
c. Frozen or broken pipes
d. Excess debris or sediment in the pipes

_____ 16. What temperature rating, temperature classification, and glass bulb color would an automatic sprinkler system need for a ceiling temperature maximum of 150°F *(66°C)?*
a. 135 to 170°F *(57°C to 77°C),* ordinary, orange or red
b. 175 to 225°F *(79°C to 107°C),* intermediate, yellow or green
c. 325 to 375°F *(163°C to 191°C),* extra high, purple
d. 500 to 575°F *(260°C to 302°C),* ultra high, black

_____ 17. What temperature rating, temperature classification, and glass bulb color would an automatic sprinkler system need for a ceiling temperature maximum of 300°F *(149°C)?*
a. 135 to 170°F *(57°C to 77°C),* ordinary, orange or red
b. 175 to 225°F *(79°C to 107°C),* intermediate, yellow or green
c. 325 to 375°F *(163°C to 191°C),* extra high, purple
d. 500 to 575°F *(260°C to 302°C),* ultra high, black

_____ 18. What sprinkler sprays a stream of water downward into a deflector that breaks the stream into a hemispherical pattern?
a. Upright
b. Sidewall
c. Pendant
d. Ceiling

_____ 19. What sprinkler extends from the side of a pipe and is used in small rooms where the branch line runs along a wall?
a. Upright
b. Sidewall
c. Pendant
d. Ceiling

_____ 20. What sprinkler sits on top of piping and sprays water into a solid deflector that breaks it into a hemispherical pattern that is redirected toward the floor?
a. Upright
b. Sidewall
c. Pendant
d. Ceiling

_____ 21. Who usually changes sprinklers when they are damaged?
a. Any firefighter with Firefighter I training
b. Firefighters who conduct the pre-incident planning
c. Fire officer or safety officer
d. Representative of the building's occupants

____ 22. Which sprinkler valve is usually located in a remote part of the sprinkler system and is used to simulate the activation of one sprinkler?
a. Post indicator valve
b. Inspector's test valve
c. Wall post indicator valve
d. Outside screw and yoke valve

____ 23. At what residual pressure should a minimum water supply be able to deliver the required volume of water to the highest sprinkler in a building?
a. 10 psi *(70 kPa)*
b. 15 psi *(100 kPa)*
c. 20 psi *(140 kPa)*
d. 25 psi *(175 kPa)*

____ 24. What capacity pumper should be used to supply water to fire department connections?
a. 250 gpm *(1 000 L/min)*
b. 500 gpm *(2 000 L/min)*
c. 750 gpm *(3 000 L/min)*
d. 1,000 gpm *(4 000 L/min)*

____ 25. What valve can be installed at the check valve and fire department connection to keep the valve and connection dry and operating properly during freezing conditions?
a. Ball drip
b. OS&Y
c. PIV
d. WPIV

____ 26. What is the required air pressure for dry-pipe systems?
a. 10 psi *(70 kPa)* below the trip pressure
b. 20 psi *(140 kPa)* below the trip pressure
c. 15 psi *(100 kPa)* above the trip pressure
d. 20 psi *(140 kPa)* above the trip pressure

____ 27. For what capacity system do NFPA standards require that a quick-opening device be installed?
a. Below 250 gallons *(1 000 L)*
b. Below 500 gallons *(2 000 L)*
c. Above 250 gallons *(1 000 L)*
d. Above 500 gallons *(2 000 L)*

____ 28. What valve discharges water from every open head connected to the system controlled by that specific valve?
a. OS&Y
b. PIVA
c. Deluge
d. Ball drip

____ 29. What NFPA standard covers residential sprinkler systems?
a. 13D
b. 1910
c. 14A
d. 1970

____ 30. Which of the following devices is *not* required for a residential sprinkler system?

a. Pressure gauge
b. System testing device
c. Ball drip valve
d. Flow detector

Identify

F. Identify the following abbreviations associated with fire detection, alarm, and suppression systems. Write the correct interpretation before each.

________________________ 1. UV

________________________ 2. IR

________________________ 3. ADA

________________________ 4. OS&Y

________________________ 5. PIVA

________________________ 6. PIV

________________________ 7. WPIV

________________________ 8. FDC

G. Identify the following fire detectors. Write the correct name below each.

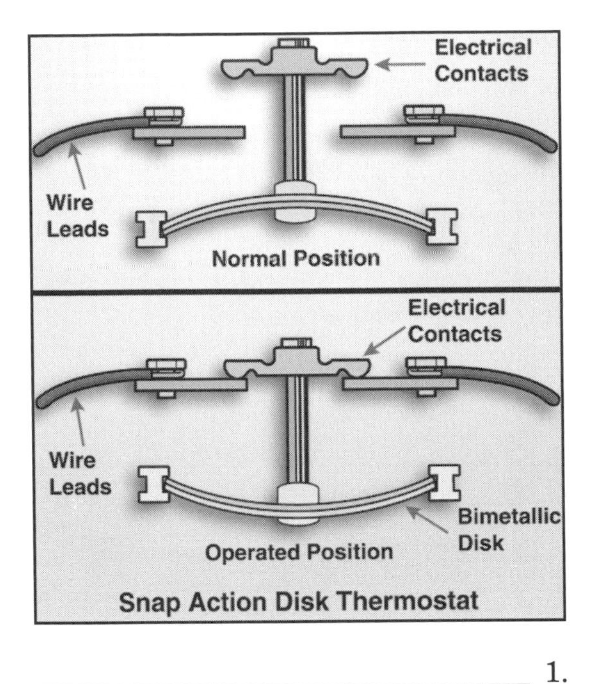

_______________ 1.

_______________ 2.

_______________ 3.

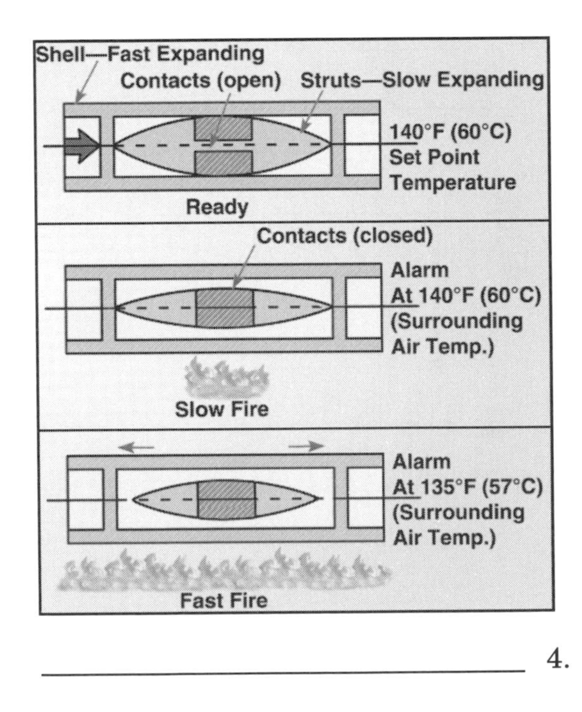

_______________ 4.

_______________ 5.

_______________ 6.

_______________ 7.

H. **Identify the following sprinklers. Write the correct name below each.**

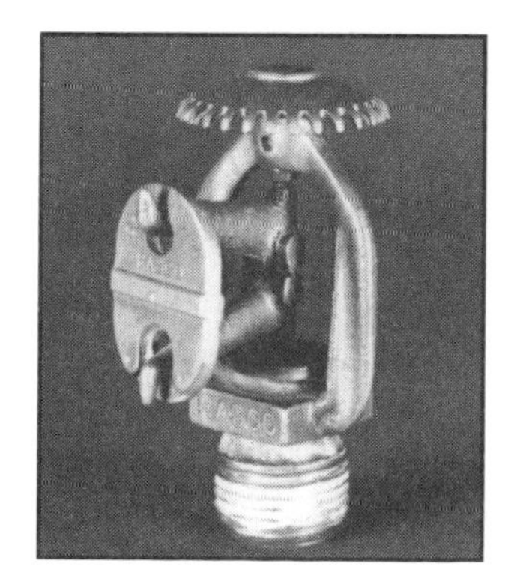

_______________ 1.

_______________ 2.

_______________ 3.

_______________ 4.

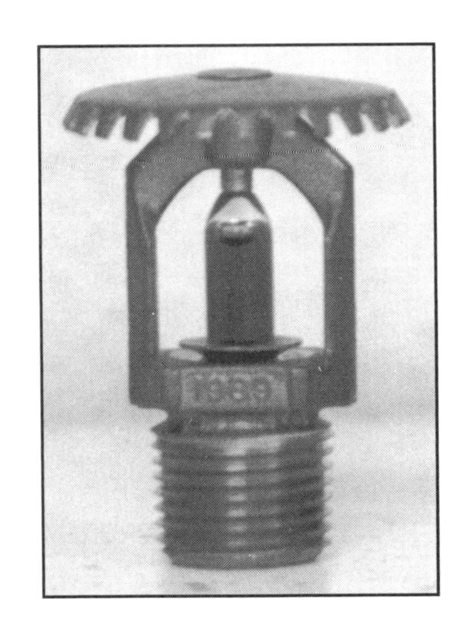

_______________ 5.

_______________ 6.

I. **Identify the following valves. Write the correct name below each.**

_______________________________ 1.

_______________________________ 2.

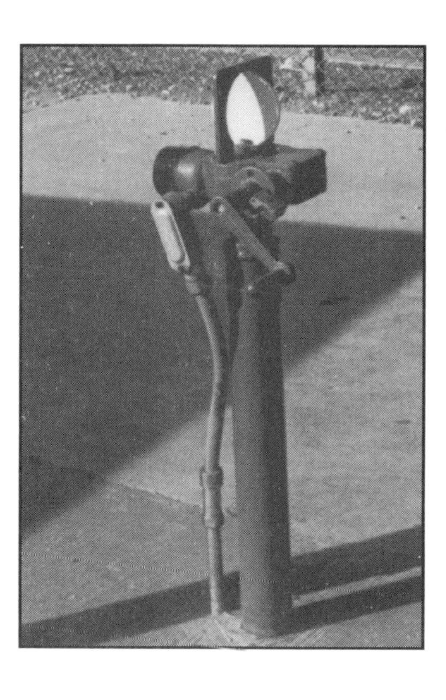

_______________________________ 3.

_______________________________ 4.

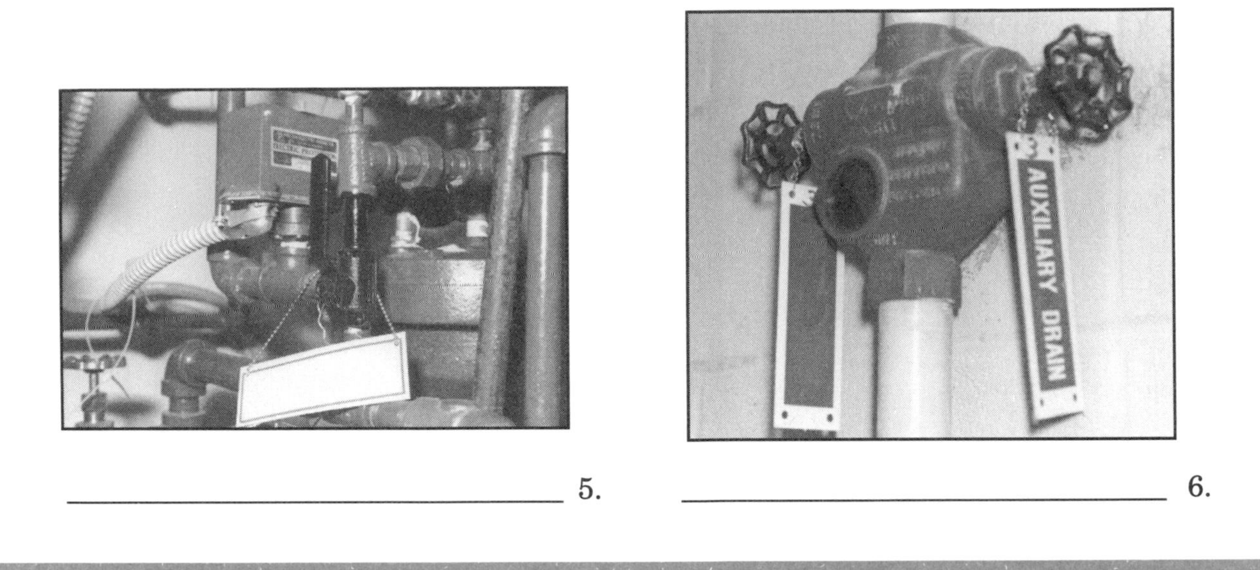

______________________ 5.　　______________________ 6.

Label

J. Label components of a complete sprinkler system. Write the correct names on the blanks next to the corresponding numbers.

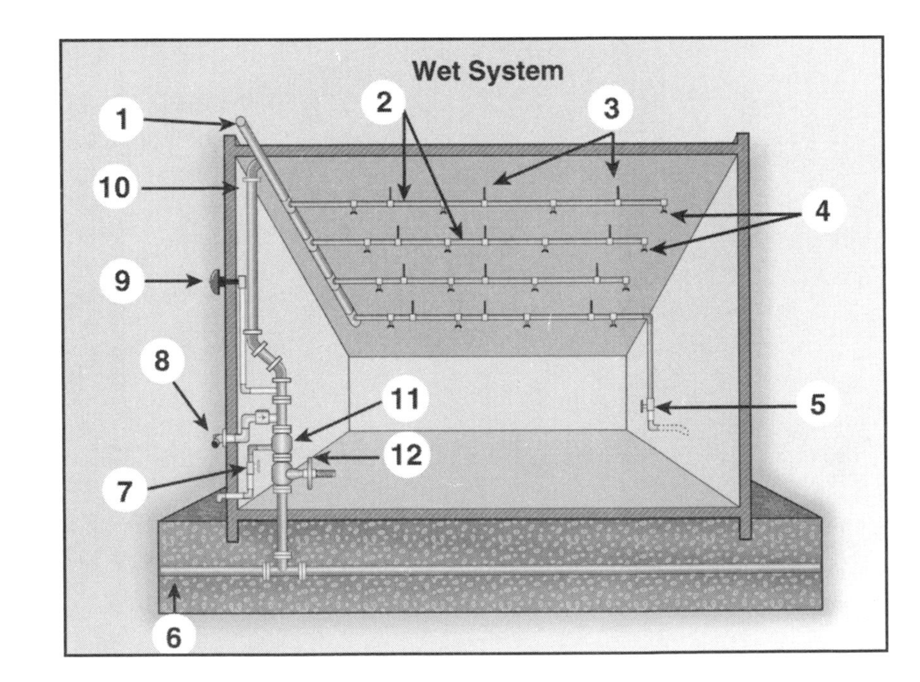

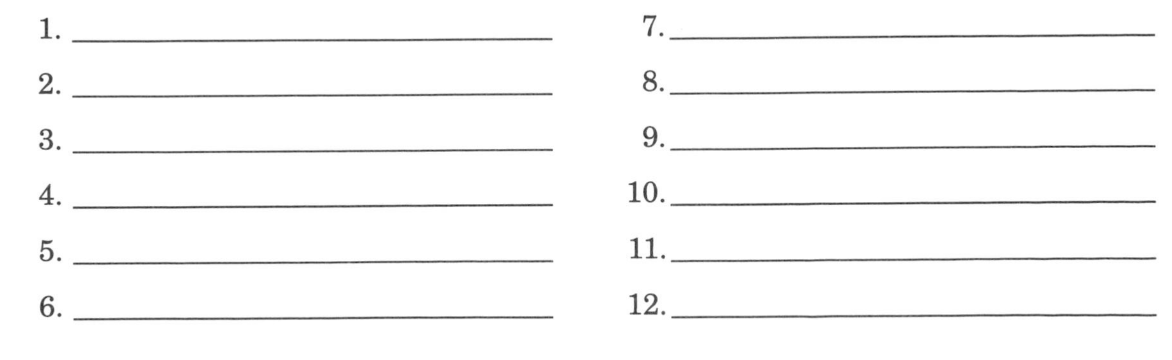

1. ______________________　　7. ______________________
2. ______________________　　8. ______________________
3. ______________________　　9. ______________________
4. ______________________　　10. ______________________
5. ______________________　　11. ______________________
6. ______________________　　12. ______________________

K. Label the components of an upright fusible-link sprinkler. Write the correct names on the blanks next to the corresponding numbers.

1. ____________________
2. ____________________
3. ____________________
4. ____________________
5. ____________________
6. ____________________

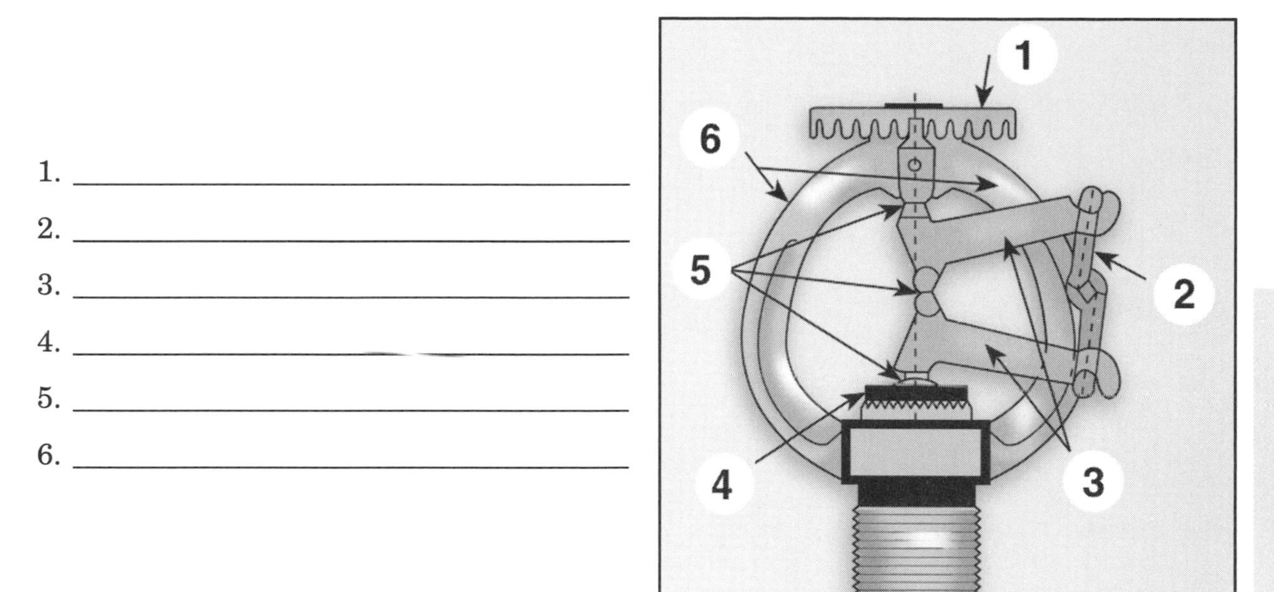

Chapter 16
Loss Control

Chapter 16 Loss Control

FIREFIGHTER I

Matching

A. Match loss control terms to their definitions. Write the appropriate letters on the blanks.

____ 1. Method of minimizing damage and providing customer service through effective mitigation and recovery efforts before, during, and after an incident

____ 2. Fire fighting technique that aids in reducing primary and secondary damage during fire fighting operations

____ 3. Operation to search for and extinguish hidden or remaining fires once the main body of fire has been extinguished

____ 4. Device used to carry debris, catch falling debris, and provide a water basin for immersing small burning objects

____ 5. Device used to protect floor coverings from being ruined by mud and grime tracked by firefighters

____ 6. Device used to remove water from basements, elevator shafts, and sumps

____ 7. Device used to remove water, dirt, and small debris from carpet, tile, and other types of floor coverings

a. Salvage
b. Water inhibitor
c. Overhaul
d. Floor runner
e. Loss control
f. Water vacuum
g. Dewatering device
h. Carryall

True/False

B. Write *True* or *False* before each of the following statements. Correct those statements that are false.

________ 1. Firefighters should wait to start salvage operations until after the fire has been knocked down.

____

____

_______ 2. Firefighters should perform overhaul operations simultaneously with fire attack.

____

____

_______ 3. Firefighters should be aware of the value of building contents vital to business survival in commercial occupancies.

____

____

_______ 4. Firefighters should adapt disposable plastic covers to traditional salvage cover folds.

____

____

_______ 5. Canvas salvage covers can be folded while wet.

____

____

_______ 6. Synthetic salvage covers do not require as much maintenance as canvas ones.

____

____

_______ 7. Firefighters may remove water from basements and elevator shafts with a fire department pumper.

____

____

_______ 8. A salvage cover rolled for a one-firefighter spread should not be carried on the shoulder.

____

____

_______ 9. Two firefighters are required to fold a salvage cover for a one-firefighter deployment.

____

____

_________ 10. Plastic sheeting is ineffective and impractical for water diversion.

____

____

_________ 11. Firefighters should preserve any possible incendiary evidence of arson.

____

____

_________ 12. Overhaul tools are job-specific and should not be used for other fire fighting operations.

____

____

_________ 13. Firefighters should begin overhaul at the point farthest from the area of actual fire involvement.

____

____

_________ 14. Insulation materials must be removed during overhaul to properly check them for hidden fire.

____

____

_________ 15. During overhaul operations, firefighters should remove large sections of wall, ceiling, and floor covering to confirm complete fire extinguishment.

____

____

_________ 16. Weight-bearing walls should not be disturbed during overhaul operations.

____

____

Multiple Choice

C. Write the letter of the best answer on the blank before each statement.

____ 1. What are the two aspects of loss control?
  a. Fire attack and salvage
  b. Evidence recovery and overhaul
  c. Salvage and evidence recovery
  d. Salvage and overhaul

____ 2. How should building contents be arranged for salvage-cover placement?
  a. Grouped in close piles
  b. Spread throughout several rooms
  c. Moved to one room with nothing stacked
  d. Piled into one corner of a room

____ 3. How can firefighters prevent furniture from absorbing water from wet flooring?
  a. Move all of the furniture to a room with wood flooring.
  b. Raise the furniture off the floor with canned goods from the kitchen.
  c. Cover the furniture completely with salvage covers.
  d. Place plywood or wooden blocks under the legs of the furniture.

____ 4. In what situation should firefighters use chutes to route water to the floor for later cleanup?
  a. When most of the furniture is inexpensive and easily replaced
  b. When the flooring is made of tile or wood
  c. When the number of salvage covers is limited
  d. When the firefighters do not have time to cover the belongings

____ 5. Which of the following statements is *not* true about synthetic covers?
  a. They are heavy and bulky.
  b. They are economical for outdoor use.
  c. They are practical for indoor use.
  d. They are easy to handle.

____ 6. What cleaning method is ordinarily required for canvas salvage covers?
  a. Scrubbing with a heavy detergent and wire brush
  b. Scrubbing with steel wool and a mild detergent
  c. Scrubbing with a soft brush and a mild detergent
  d. Scrubbing with a broom and a hose stream

____ 7. How should firefighters mark holes in canvas salvage covers?
   a. Masking tape
   b. Electric tape
   c. Duct tape
   d. Chalk

____ 8. Who or what dictates on which apparatus the salvage equipment is carried and who performs the primary salvage operations on the fire scene?
   a. Incident commander
   b. NFPA standards
   c. Fire department SOPs
   d. Fire safety officer

____ 9. Which of the following tools would *not* be used to stop the flow of water from an open sprinkler?
   a. Sprinkler tongs
   b. Sprinkler wrench
   c. Sprinkler stopper
   d. Sprinkler wedge

____ 10. What is the most convenient way to carry a salvage cover folded for a two-firefighter spread?
   a. On the shoulder with the open edges next to the neck
   b. On the shoulder with the open edges away from the neck
   c. Under the arm with the open edges facing downward
   d. Under the arm with the open edges facing upward

____ 11. What should firefighters use to cover doors and windows that have been broken or removed?
   a. Trash bags
   b. Thin plastic
   c. Cardboard
   d. Plywood

____ 12. Which of the following tools should be used to open ceilings to check for fire extension?
   a. Bale hook
   b. Axe
   c. Pike pole
   d. Claw hammer

____ 13. Which of the following tools should be used to open walls and floors to check for fire extension?
   a. Bale hook
   b. Axe
   c. Pike pole
   d. Claw hammer

____ 14. When can a firefighter forego wearing SCBA during overhaul operations?
   a. When the air is visually clear and no toxic odor is detected
   b. When no flames and very little smoke are present
   c. When the structure has burned completely to the ground and no walls or ceilings are still standing
   d. When the air has been proven safe through testing

____ 15. Which of the following procedures should be followed during overhaul operations?
a. Use a 1-inch *(25 mm)* attack line.
b. Never use a water fire extinguisher.
c. Keep one attack line available.
d. Never use booster hoses.

____ 16. Which of the following methods should *not* be used to detect hidden fires?
a. Smell for acrid or pungent odors.
b. Feel walls and floors for heat.
c. Listen for popping, cracking, or hissing sounds.
d. Look for cracked plaster or rippled wallpaper.

____ 17. How should firefighters extinguish small burning objects found during overhaul?
a. Drench the object with a fire stream.
b. Plunge the object into a sink or tub of water.
c. Stomp on the object to smother the fire.
d. Cover the object with a firefighter's coat.

____ 18. How should firefighters ensure fire extinguishment in bales of alfalfa?
a. Drench the bale with Class A foam.
b. Wet the top of the bale and roll the bale to wet the bottom.
c. Break the bale apart.
d. Soak the bale with water and watch it for at least an hour.

Chapter 17
Protecting Evidence for Fire Cause Determination

Chapter 17 Protecting Evidence for Fire Cause Determiniation

FIREFIGHTER I

True/False

A. Write *True* or *False* before each of the following statements. Correct those statements that are false.

_______ 1. Timeliness is the most important part of evidence discovery.

____

____

_______ 2. Firefighters should collaborate in teams of two or three to write a chronological account of the circumstances they observe at a fire scene.

____

____

_______ 3. Hearsay cannot be used in court; therefore, firefighters should not report hearsay to an investigator.

____

____

_______ 4. Firefighters should delay thorough salvage and overhaul work until the area of origin has been determined.

____

____

_______ 5. Firefighters should gather evidence and store it in a safe place.

____

____

_______ 6. Firefighters who handle or procure evidence become a link in the chain of custody for that evidence.

____

____

_____ 7. Firefighters may remove debris around evidence as long as they do not disturb the evidence.

____

____

Multiple Choice

B. Write the letter of the best answer on the blank before each statement.

___ 1. When does a firefighter's responsibility for gathering fire cause information begin?
   - a. When the alarm is received
   - b. After completing rescue and extrication
   - c. During the fire attack
   - d. During overhaul and salvage

___ 2. What fire scene operations are pivotal in determining fire cause?
   - a. Fire attack and ventilation
   - b. Rescue and extrication
   - c. Overhaul and salvage
   - d. Ventilation and fire stream selection

___ 3. How can firefighters best preserve human footprints and tire marks?
   - a. Place salvage covers over them.
   - b. Park apparatus over them.
   - c. Cordon off the area.
   - d. Place a box over them.

___ 4. What should firefighters do with completely or partially burned papers found in a furnace, stove, or fireplace?
   - a. Remove them and place them in plastic evidence bags.
   - b. Douse them with water to prevent further fire or heat damage.
   - c. Close dampers and other openings that may disturb them.
   - d. Make note of them and remove them with the other debris.

___ 5. Where should firefighters dump the debris from a burned structure?
   - a. In the backyard or alley
   - b. In the front yard or street
   - c. In a nearby parking lot
   - d. In the closest sewer system

Case Study

C. Read the following scenarios and answer the questions that follow each. Answers can be in question or statement format. Provide at least two answers for each scenario.

1. Firefighters are dispatched at 2 a.m. to a fire in a commercial building still under construction. As the first engine approaches the scene, firefighters see a man in work clothes walking away from the building and toward the approaching fire apparatus. Firefighters can see flames coming from the front right corner of the building and from the left side of the roof.

   Based on this scenario, what key observations/questions should firefighters note for a possible fire cause investigation?

   ____

   ____

   ____

   ____

   ____

   ____

   ____

2. Upon arriving at a residential fire on a busy street in late summer, firefighters find a fire in the garage of a single-story house. The garage door is open. Several vehicles are parked along the curb, and one of them is blocking the hydrant. Police locate the owner, and she moves her vehicle so that the fire apparatus can connect to the fire hydrant.

   The fire appears to be located only in the garage. Several cans of paint and an assortment of carpentry tools are stored on the shelves in the garage. A small space heater is located near a workbench, and the workbench and floor are both free of dust and wood shavings.

   Based on this scenario, what key observations/questions should firefighters note for a possible fire cause investigation?

   ____

   ____

   ____

   ____

   ____

   ____

   ____

3. Firefighters arrive at a grocery store fully involved in fire. People in street clothes and others wearing smocks are standing in small groups in the parking lot watching the fire. Several vehicles are pulling into the parking lot, and a few are leaving. The loading docks behind the store are all closed except one that has a semitrailer backed to it.

   During salvage and overhaul operations, firefighters carefully step over rubble and move obstructions to reach the office area to search for valuable documents. At the store owner's request, firefighters search for and recover a "fireproof" locked box containing the store's insurance policy.

   Based on this scenario, what key observations/questions should firefighters note for a possible fire cause investigation?

   ____

   ____

   ____

   ____

   ____

   ____

   ____

4. Firefighters arrive at the scene of an early-morning fire in a vacant office building. Old furniture in the hallway hinders the initial attack. Several rekindles occur. Firefighters find broken glass near the seat of the fire. Although the building has a fire alarm system, a passing motorist saw the fire and made the first call to 9-1-1.

   Based on this scenario, what key observations/questions should firefighters note for a possible fire cause investigation?

   ____

   ____

   ____

   ____

   ____

   ____

   ____

5. Shortly after lunch, firefighters respond to a report of smoke coming from a residential home. Upon arrival, firefighters notice smoke venting from a side window. After assessing the scene, they stretch an attack line to the front door. They make entry and follow a path of charred carpet to a bedroom where they attack a mattress fire. Although the fire is still in its incipient stage, a large pile of burning paper on the mattress scatters when hit with water because some debris from the fire hydrant comes through the hoseline.

   Based on this scenario, what key observations/questions should firefighters note for a possible fire cause investigation?

FIREFIGHTER II

True/False

A. Write *True* or *False* before each of the following statements. Correct those statements that are false.

_______ 1. Even in the midst of fire fighting operations, firefighters should be careful to protect evidence.

_______ 2. Investigators are usually present to interview occupants and witnesses while firefighters perform fire fighting operations.

_______ 3. Firefighters should question any potential arson suspect.

_______ 4. The fire department is authorized to bar access to any building for a reasonable length of time even after fire suppression is terminated.

_______ 5. Firefighters must follow NFPA guidelines when cordoning off the area at a fire scene.

_______ 6. Firefighters should obtain identification and an explanation from anyone who attempts to cross into a cordoned area even if the person is a member of the news media.

_______ 7. Firefighters must obtain a search warrant before removing any evidence from a fire scene during overhaul operations.

____

____

_______ 8. If incendiary evidence is found, at least one fire service representative should remain at the fire scene until an arson investigator arrives.

____

____

Multiple Choice

B. Write the letter of the best answer on the blank before each statement.

___ 1. Who should *not* collect evidence at a fire scene?

a. Firefighter I
b. Firefighter II
c. Fire officer
d. Fire investigator

___ 2. In most jurisdictions, who has the legal responsibility for determining the cause of a fire?

a. Fire officer
b. Incident commander
c. Safety officer
d. Fire chief

___ 3. Who is usually responsible for carrying fire cause investigations beyond the level of the fire company?

a. City mayor
b. Fire marshal
c. Governor's council
d. Local police

___ 4. How should firefighters reply to news media questions concerning the cause of a fire?

a. *"No comment."*
b. *"We do not suspect arson at this time."*
c. *"The fire is under investigation."*
d. *"Please refer all questions to the district attorney."*

___ 5. To whom should fire personnel make statements at a fire scene?

a. Incident commander
b. Reporters
c. Owners or occupants of the structure
d. Fire investigator

____ 6. What should firefighters do with evidence if an investigator is not immediately available?

a. Mark, tag, and photograph the evidence.

b. Collect the evidence and take it to the police department.

c. Gather the evidence into a localized area before removing debris from the structure.

d. Leave the evidence where it was found and return to the scene when the investigator arrives.

____ 7. Which of the following items does *not* have to be recorded when someone enters a fire scene during an investigation?

a. Name

b. Relationship to the owner

c. Length of time in the area

d. Description of items taken from the area

____ 8. According to the *Michigan vs. Tyler* U.S. Supreme Court decision, when are fire personnel and/or arson investigators required to get a warrant to enter a fire scene?

a. Any time that arson is suspected

b. When the investigation takes longer than 24 hours

c. If the occupant or owner requests a warrant

d. If firefighters have left the premises and want to reenter

Chapter 18
Fire Department Communications

Chapter 18 Fire Department Communications

FIREFIGHTER I

True/False

A. Write *True* or *False* before each of the following statements. Correct those statements that are false.

_______ 1. Telecommunicators are not expected to know where specific emergency resources are located in relation to reported incidents.

____

____

_______ 2. Telecommunicators must evaluate a caller's worthiness for assistance.

____

____

_______ 3. Computers and electronic displays have negated the need for telecommunicators to read a map.

____

____

_______ 4. Automatic vehicle locating systems have reduced the need for telecommunicators to use maps.

____

____

_______ 5. TTY and TDD are devices that permit hearing- and speech-impaired people to communicate with the fire department.

____

____

_______ 6. A computer-generated document must be converted to hard copy before it can be transmitted to a stand-alone fax machine.

____

____

_______ 7. Telecommunicators should get the same information from a firefighter reporting an emergency over the radio as they would get from someone calling in an emergency on the telephone.

____

____

_______ 8. Staffed fire stations need loud audible devices and bright lights to alert personnel of an emergency.

____

____

_______ 9. Pagers are more efficient than home monitors for alerting off-duty or volunteer firefighters of an emergency call.

____

____

_______ 10. Alerting fire personnel with sirens, whistles, and air horns is most commonly employed in small communities.

____

____

_______ 11. Fire department radio communication in the United States is licensed by the Federal Communications Commission.

____

____

_______ 12. Firefighters should send only a limited number of personal messages over the designated fire department radio channel.

____

____

_______ 13. All fire service personnel should provide adequate justification for requests made over the radio.

____

____

_____ 14. Fire personnel may need to use several radio channels during large incidents to enable clear and timely information exchange.

____

____

Multiple Choice

B. Write the letter of the best answer on the blank before each statement.

_____ 1. What is the generally accepted time period to effect a dispatch?

a. 1 minute
b. 2 to 3 minutes
c. 4 minutes
d. 5 minutes

_____ 2. With whom should the telecommunicator stay in contact during an incident?

a. Any driver/operator on the scene
b. Any firefighter with Firefighter II training
c. Fire chief
d. Incident commander

_____ 3. How should a telecommunicator deal with someone who dials 9-1-1 for a nonemergency matter?

a. Ask the caller to hang up and call Directory Assistance.
b. Inform the caller that the communications center accepts only phone calls relating to an emergency.
c. Transfer or refer the caller to an agency that can answer the question.
d. Report the caller's name and address to law enforcement officials.

_____ 4. According to the appendix of NFPA 1061, when may firefighters perform the duties of a telecommunicator?

a. When they meet the requirements of NFPA 1061
b. When no certified telecommunicator is available
c. When the telecommunications center has an overflow workload
d. After they meet the requirements for Firefighter I certification

_____ 5. How does an AVL system help telecommunicators?

a. The system displays the phone number and location of a caller on a computer monitor.
b. The system displays the location of a fire department unit on a map.
c. The system automatically calls back any disconnected caller.
d. The system records all incoming calls.

____ 6. What device is the most widely used system for transmitting fire alarms?

a. Telephone fire alarm box
b. Radio fire alarm box
c. Direct telephone line
d. Public telephone line

____ 7. Who has the ability to monitor radio transmissions broadcast from a telecommunications center?

a. Only law enforcement officials and fire service personnel
b. Only fire service personnel
c. Anyone with a receiver, including the news media and the public
d. Only authorized fire service, law enforcement, and government personnel

____ 8. How does computer-aided dispatch affect telecommunications?

a. It eliminates the need for telecommunicators.
b. It enables dispatchers to handle a greater volume of calls.
c. It replaces the need for radio calls between telecommunicators and responding units.
d. It decreases the number of false alarms.

____ 9. What voice-recording documentation system may omit the beginning of a transmission?

a. Intermittent
b. Continuous
c. Radio log
d. Supplemental

____ 10. What documentation system is manually entered onto paper?

a. Intermittent
b. Continuous
c. Radio log
d. Supplemental

____ 11. What basic 9-1-1 feature allows a telecommunicator to maintain access to a caller's phone line by *not* hanging up or disconnecting?

a. System control
b. Forced disconnect
c. Called party hold
d. Ringback

____ 12. What basic 9-1-1 feature allows a telecommunicator to dial back to a caller's phone after the caller terminates the connection?

a. System control
b. Forced disconnect
c. Called party hold
d. Ringback

____ 13. What basic 9-1-1 feature allows the telecommunicator to drop a call out of the system to open the line for another caller?

a. System control
b. Forced disconnect
c. Called party hold
d. Ringback

____ 14. What E-9-1-1 feature displays the calling party's phone number on a display screen at the telecommunicator's position?

a. ANI
b. CAD
c. ALI
d. TDD

____ 15. What E-9-1-1 feature displays the calling party's location, phone number, directions to the location, and other information about the address?

a. ANI
b. CAD
c. ALI
d. TDD

____ 16. What CB radio channel is the universal frequency for reporting emergencies?

a. 19
b. 17
c. 11
d. 9

____ 17. Instead of a callback number, what information should a telecommunicator get from a CB radio caller?

a. Caller's preferred channel
b. Caller's radio "handle"
c. Location of the emergency
d. Home phone number

____ 18. Which alarm system is limited in that the only information transmitted is the location?

a. Telephone fire alarm box
b. Direct telephone line
c. Wired telegraph circuit box
d. Citizen's band radio

____ 19. What system of transmitting dispatch information reduces overall response time?

a. Sending information over the radio before the fire apparatus leave the station
b. Looking up information on transparencies in the fire apparatus
c. Retrieving information from microfiche in the fire apparatus
d. "Pre-alerting" the fire station while researching dispatch information

____ 20. Which of the following methods should a telecommunicator use to effectively extract information from reluctant, scared, and upset victims/witnesses?

a. Keep vocal pitch at midrange.
b. Emphasize each word with accentuated articulation.
c. Speak very slowly.
d. Use dialects or regionalisms as necessary.

____ 21. How should a firefighter at an emergency scene confirm receipt of a transmitted communication?

a. Ask the sender to repeat the communication.
b. Reply, *"10-4."*
c. Forward the communication to the incident commander.
d. Repeat the communication to the sender.

____ 22. Where and how should firefighters hold a radio/microphone while transmitting a message?

a. At least 5 inches *(125 mm)* from the mouth and in a vertical position

b. About 1 to 2 inches *(25 mm to 50 mm)* from the mouth and at a 45-degree angle

c. Within 3 inches *(75 mm)* of the mouth and parallel to the head

d. Between 3 and 5 inches *(75 mm and 125 mm)* from the face and at any angle

____ 23. What report should be given over the radio by the first arriving companies to provide information about conditions at the incident?

a. Incident report

b. Radio log

c. Size-up

d. Pre-alert

____ 24. Which of the following updates should be sent to the telecommunications center in a progress report?

a. Change in command location

b. Number of personnel and apparatus at the scene

c. Names and certification levels of team leaders

d. Names and addresses of victims and missing persons

____ 25. Which of the following incidents would *least* likely require the use of a tactical channel?

a. Apartment building fire

b. Vehicle fire

c. Building collapse

d. Wildland fire

____ 26. Who is responsible for assigning an operational frequency for the management of an incident?

a. Incident commander

b. First arriving fire company

c. Telecommunicator

d. Fire officer

____ 27. Who is most likely to hear a weak radio signal from a firefighter using a portable or mobile radio at an emergency scene?

a. Incident commander

b. Personnel en route to the scene

c. On-scene personnel

d. Telecommunicator

____ 28. How should personnel from Central 1 company alert the telecommunicator when they have urgent broadcast messages?

a. *"Dispatch, this is Central 1, please make an emergency broadcast!"*

b. *"Dispatch, this is Central 1, emergency traffic!"*

c. *"Dispatch, this is Central 1, I have an urgent message for all personnel on the scene!"*

d. *"Dispatch, this is Central 1, urgent message, please broadcast!"*

Identify

C. **Identify the following abbreviations associated with fire department communications. Write the correct interpretation before each.**

______________________ 1. CAD

______________________ 2. AVL

______________________ 3. TDD

______________________ 4. TTY

______________________ 5. ANI

______________________ 6. ALI

______________________ 7. CB

______________________ 8. FCC

FIREFIGHTER II

True/False

A. Write *True* or *False* before each of the following statements. Correct those statements that are false.

_______ 1. All firefighters need to know the local procedure for requesting additional alarms during an emergency.

____

____

_______ 2. Incident reports are confidential.

____

____

_______ 3. Incident reports should be handwritten.

____

____

Multiple Choice

B. Write the letter of the best answer on the blank before each statement.

____ 1. Who normally orders multiple alarms or additional responses?

a. Only the incident commander
b. Any firefighter at the scene
c. Any company officer at the scene
d. Any chief officer at the scene

____ 2. What radio communication option can be used during a large fire to reduce the load on the telecommunications center?

a. Limit radio airtime for personnel at the scene.
b. Limit number of radio transmissions per responding unit.
c. Use a mobile, radio-equipped, command vehicle.
d. Restrict number of radio transmissions from the scene.

____ 3. According to NFPA 902, what report must be completed every time a fire unit responds to an incident?

a. Status report
b. Radio log
c. Incident report
d. Progress report

____ 4. How does NFIRS transfer data from each state to the federal database?

a. Internet
b. Fax
c. Network hubs
d. Priority mail

Identify

C. Identify the following abbreviations associated with fire department communications. Write the correct interpretation before each.

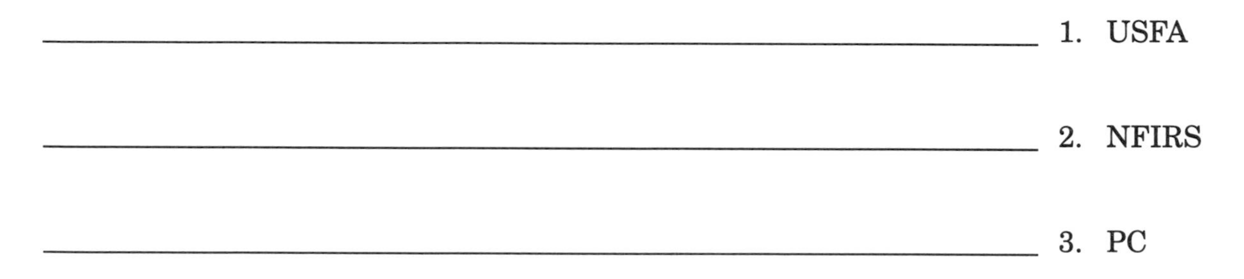

________________________________ 1. USFA

________________________________ 2. NFIRS

________________________________ 3. PC

D. Identify the information that should be included in an incident report. Place an *X* next to each correct answer.

________ 1. Incident number

________ 2. Names of fire service personnel at the scene

________ 3. Primary use, number of stories, and construction type of structure

________ 4. Names and ages of people injured or fatally wounded

________ 5. Number of personnel who responded

________ 6. How and where the fire started

________ 7. Estimated cost of damage

________ 8. How the emergency was reported (9-1-1, walk-in, radio)

________ 9. Name(s) of the incident commander(s)

________ 10. Amount of extinguishing agent used and/or a list of medical supplies used

Chapter 19
Fire Prevention and Public Fire Education

Chapter 19 Fire Prevention and Public Fire Education

FIREFIGHTER I

Matching

A. Match to their definitions terms associated with fire prevention and public fire education. Write the appropriate letters on the blanks.

____ 1. Activities conducted by firefighters; designed to discover potential fire hazards and to inform building occupants and owners of unsafe conditions

____ 2. Reports that contain the documented fire history of a community

____ 3. Investigations conducted by trained professionals who have met the objectives of NFPA 1031; designed to discover potential fire hazards and to communicate unsafe conditions to building occupants and owners

____ 4. Conditions that encourage a fire to start or increase the extent or severity of a fire

a. Pre-incident plans
b. Fire hazards
c. Fire safety inspections
d. Fire safety surveys
e. Fire incident records

B. Match fire hazard types to their definitions. Write the appropriate letters on the blanks.

____ 1. Condition that is prevalent in almost all occupancies and encourages a fire to start

____ 2. Common hazard caused by the unsafe acts of individuals

____ 3. Condition that arises as a result of processes or operations that are characteristic of an individual occupancy

____ 4. Facility in which there is a great potential for life or property loss from a fire

a. Common fire hazard
b. Fixed fire hazard
c. Special fire hazard
d. Personal fire hazard
e. Target fire hazard

True/False

C. Write *True* or *False* before each of the following statements. Correct those statements that are false.

_______ 1. Residential fire safety surveys are usually part of a fire department's public awareness and education programs.

_______ 2. Firefighters are authorized to perform residential fire safety surveys despite protests from an owner/occupant.

_______ 3. Children under the age of ten should not be included in fire safety education programs.

_______ 4. During fire and life safety presentations, firefighters should encourage the audience to demonstrate life safety activities.

_______ 5. Firefighters conducting fire and life safety presentations should insist that audience members make safe practices a way of life.

_______ 6. It is a proven fact that citizens can safely escape during home fire emergencies with proper preparation and practice.

_______ 7. Test buttons on smoke detectors check the detector's sensitivity to smoke.

_______ 8. Firefighters should dress appropriately and conduct only productive activities while citizens are in the fire station.

____

____

_______ 9. After a fire station tour, firefighters may allow visitors to explore safe areas of the station on their own.

____

____

Multiple Choice

D. Write the letter of the best answer on the blank before each statement.

____ 1. Which of the following gases is the most difficult fire hazard to control?
- a. Liquefied petroleum gas
- b. Natural gas
- c. Compressed natural gas
- d. Oxygen

____ 2. What fire hazard category includes the misuse of fumigation substances and flammable or combustible liquids?
- a. Fixed
- b. Common
- c. Special
- d. Target

____ 3. Which of the following special hazards is generally found in commercial occupancies?
- a. High-piled storage of combustible materials
- b. Poor housekeeping and improper storage of packing materials
- c. Existence of party walls, common attics, or cocklofts
- d. Insufficient, blocked, or locked exits

____ 4. Which of the following special hazards is generally found in a manufacturing occupancy?
- a. High-piled storage of combustible materials
- b. Poor housekeeping and improper storage of packing materials
- c. Existence of party walls, common attics, or cocklofts
- d. Insufficient, blocked, or locked exits

____ 5. Which of the following special hazards is generally found in a public assembly occupancy?
- a. High-piled storage of combustible materials
- b. Poor housekeeping and improper storage of packing materials
- c. Existence of party walls, common attics, or cocklofts
- d. Insufficient, blocked, or locked exits

____ 6. Which of the following items should a firefighter take to a survey site?

a. Standard plan symbols
b. Reference books
c. Drawing scales
d. Code and inspection manuals

____ 7. What percentage of all fires and the vast majority of civilian casualties occur in residences?

a. 40
b. 50
c. 70
d. 85

____ 8. For which of the following residential dwellings do fire codes require inspections?

a. One-family farm house
b. One-family multilevel house
c. Two-family duplex
d. Four-family quadplex

____ 9. Which of the following objectives should *not* be part of a residential fire safety survey?

a. Pointing out code violations
b. Improving life safety conditions
c. Preventing accidental fires
d. Helping the occupant improve existing conditions

____ 10. How do firefighters who conduct residential safety surveys benefit their local fire departments?

a. By enforcing fire codes
b. By creating positive public relations
c. By establishing fire department authority
d. By soliciting donations

____ 11. Which of the following guidelines should firefighters use when conducting residential safety surveys?

a. Maintain a serious and authoritative attitude.
b. Instruct the occupant to make corrections when hazards are found.
c. Show other residential surveys to the occupant for comparison.
d. Compliment the occupant when favorable conditions are found.

____ 12. When firefighters are canvassing a neighborhood, they should *not* leave educational materials for absent residents ____.

a. Between an interior door and a screen door
b. Partially beneath a doormat
c. In a mailbox
d. In a bag hanging on the doorknob

____ 13. Firefighters should strongly recommend that occupants place one smoke detector ____.

a. In every bedroom and one at every level of the living unit
b. In every hallway and one in the kitchen
c. In the living area and one in the kitchen
d. Outside each of the bedrooms and one in the living area

____ 14. Where is the best place to mount smoke detectors?

a. In the dead air space
b. At eye level on the wall
c. On the wall within 2 inches *(50 mm)* of the ceiling
d. On the ceiling

____ 15. Which of the following safety guidelines should be followed when firefighters have visitors at the fire station?

a. Place a firefighter at each corner of apparatus.
b. Take visitors on elevated platforms two at a time.
c. Supervise children climbing on aerial ladders.
d. Blow sirens for children if they all agree.

Identify

E. Identify the following abbreviations associated with fire prevention and public education. Write the correct interpretation before each.

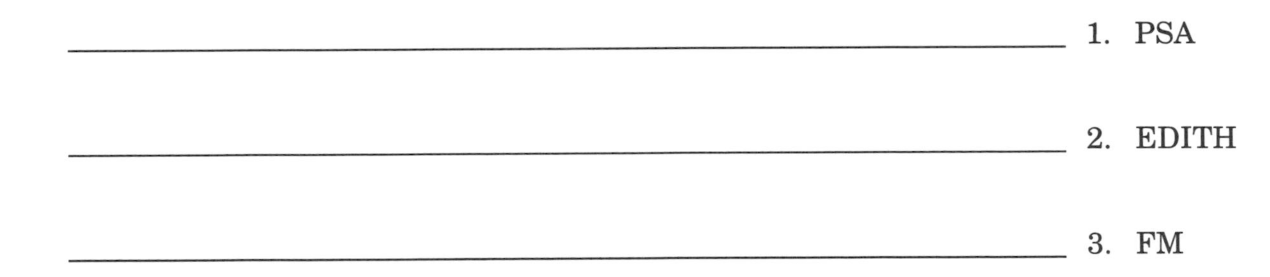

________________________________ 1. PSA

________________________________ 2. EDITH

________________________________ 3. FM

F. Identify the home safety rules that firefighters should communicate during fire and life safety presentations. Place an *X* before each correct rule.

________ 1. Keep doors to bedrooms open while sleeping.

________ 2. Have two (or more) escape routes from every room.

________ 3. Never open windows during a fire; this action could create improper ventilation.

________ 4. Train children properly if they are expected to use fire escape ladders.

_________ 5. Alert other family members by crawling from room to room.

_________ 6. Roll out of bed onto the floor if awakened by a smoke detector alarm.

_________ 7. Call the fire department before doing anything else.

_________ 8. Stay low at all times.

_________ 9. If the floor is warm to the touch, leave the room through the hallway.

_________ 10. Never go back into the house after escaping.

Case Study

G. Read the following scenarios and answer the questions that follow each. Provide at least five answers for each scenario. Note good fire safety practices as well as possible hazards.

1. Firefighters enter a one-story residential home with a fireplace and a two-car attached garage to conduct a residential fire safety survey. The living area contains a couch, two chairs, and a coffee table. Magazines and newspapers are stored under the table. Each of the three bedrooms, the hallway, the living area, and the kitchen have smoke detectors installed on the ceilings. The utility closet in the kitchen houses the hot water tank, washer, dryer, household cleaning agents, and rags. The occupants use one side of the garage as a woodworking shop. The garage does not have air-conditioning or heating vents, so the occupants have outfitted the area with a water cooler and an electric space heater.

   Based on this scenario, what key observations/questions should firefighters discuss with occupants?

   ____

   ____

   ____

   ____

   ____

   ____

   ____

2. Firefighters examine the exterior of a residential home with a fireplace during a residential fire safety survey. The house is built on a crawl space and has an elevated porch. The residents are concerned with security and have installed window bars. The shrubbery around the porch is trimmed. Beneath the porch are several cans of paint and old paint brushes. The occupant keeps a 50-gallon *(200 L)*

drum for burning trash behind the house. Near the back of the property is a padlocked shed where the lawnmower, extra gasoline, fertilizers, and pesticides are located.

Based on this scenario, what key observations/questions should firefighters discuss with occupants?

____

____

____

____

____

____

____

FIREFIGHTER II

True/False

A. Write *True* or *False* before each of the following statements. Correct those statements that are false.

_______ 1. During a pre-incident survey, firefighters should establish an authoritative relationship with the building owner.

____

____

_______ 2. Firefighters should conduct a pre-incident survey without the occupant or a representative acting as a chaperone.

____

____

_______ 3. Firefighters can perform much of their haz mat identification training at local commercial and industrial facilities.

____

____

_______ 4. According to NFPA 704, firefighters are required to affix a haz mat marking system on the outside of any structure housing such materials.

____

____

_______ 5. If the floor plan used on a previous survey is available, firefighters may use that floor plan to speed the pre-incident survey process.

____

____

Multiple Choice

B. Write the letter of the best answer on the blank before each statement.

____ 1. When should firefighters survey the exterior of the building?
a. Prior to setting up the meeting with the owner
b. After conducting the interior survey
c. Before the initial meeting with the owner
d. After the initial meeting with the owner

____ 2. Which of the following exterior conditions would firefighters *not* record on a survey?
a. Exact dimensions of the parking lots
b. Location of standpipes
c. General housekeeping of the area
d. Height of adjacent exposures

____ 3. In what order should firefighters survey the interior of a building?
a. Start with the offices, move to the production area, and finally go to the basement or roof.
b. Start on the first floor with one team advancing to the roof and the other proceeding to the basement.
c. Start on the roof and proceed systematically to the basement.
d. Start with the haz mat areas and continue systematically to the least hazardous areas.

____ 4. What should firefighters conducting a pre-incident survey do when barred from secret areas of a building?
a. Omit those areas from the survey.
b. Highlight those areas on the sketches.
c. Bring in law enforcement officials to gain access.
d. Report the incident to the fire marshal.

____ 5. What sketch shows the general arrangement of a property with respect to streets and other buildings?
a. Floor plan
b. Plot plan
c. Sectional elevation sketch
d. Exposure survey

____ 6. What sketch should be used to show mezzanines or balconies?
a. Floor plan
b. Plot plan
c. Sectional elevation sketch
d. Exposure survey

____ 7. How do firefighters prepare pre-incident survey sketches and notes for future reference and classroom study?
   a. Redraw the sketches to scale.
   b. Laminate all survey documents.
   c. Convert the survey documents to microfiche.
   d. Make copies of the survey for local and state law enforcement records.

____ 8. After pre-incident survey sketches and notes are complete, how should firefighters conclude the survey process?
   a. Evaluate the information and send a formal letter to the owner stating the current conditions and any possible recommendations.
   b. Contact the owner by phone with the results of the survey.
   c. Submit the findings to the fire prevention officer who will then contact the owner with the results.
   d. Meet with the person in authority to explain the findings before leaving the premises.

Identify

C. Identify the following abbreviations associated with fire prevention and public fire education. Write the correct interpretation before each.

____ 1. haz mat

____ 2. GIS

Label

D. Identify the following standard map symbols. Write the correct interpretation next to each. Some symbols may have more than one correct answer.

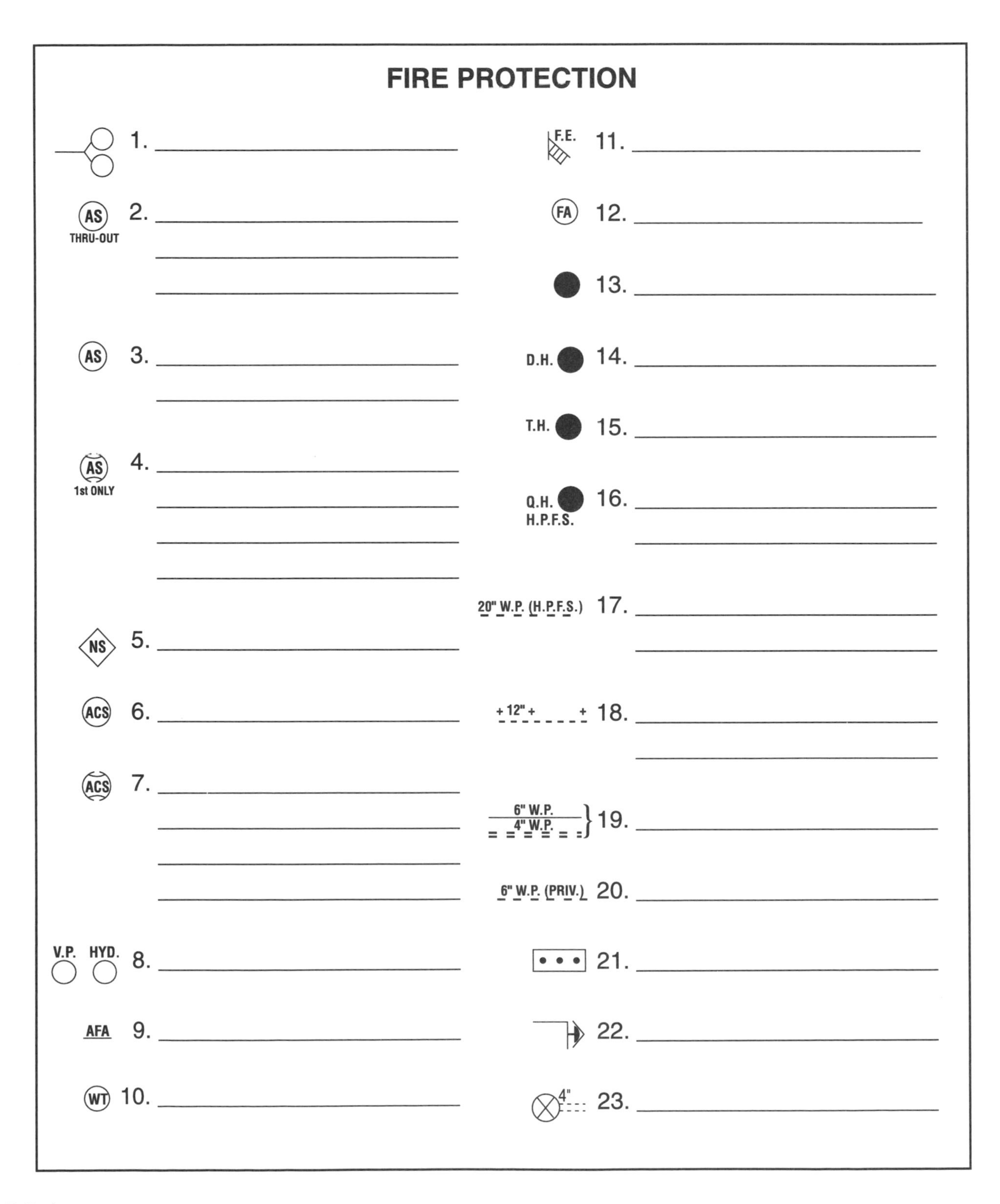

VERTICAL OPENINGS

Label	No.	Label	No.
	1. ______	CBET	8. ______
3	2. ______	TESC	9. ______
WG	3. ______	BE	10. ______
E	4. ______	H	11. ______
FE	5. ______	HT	12. ______
ET	6. ______	H (B. to 1)	13. ______
ESC	7. ______	STAIRS	14. ______

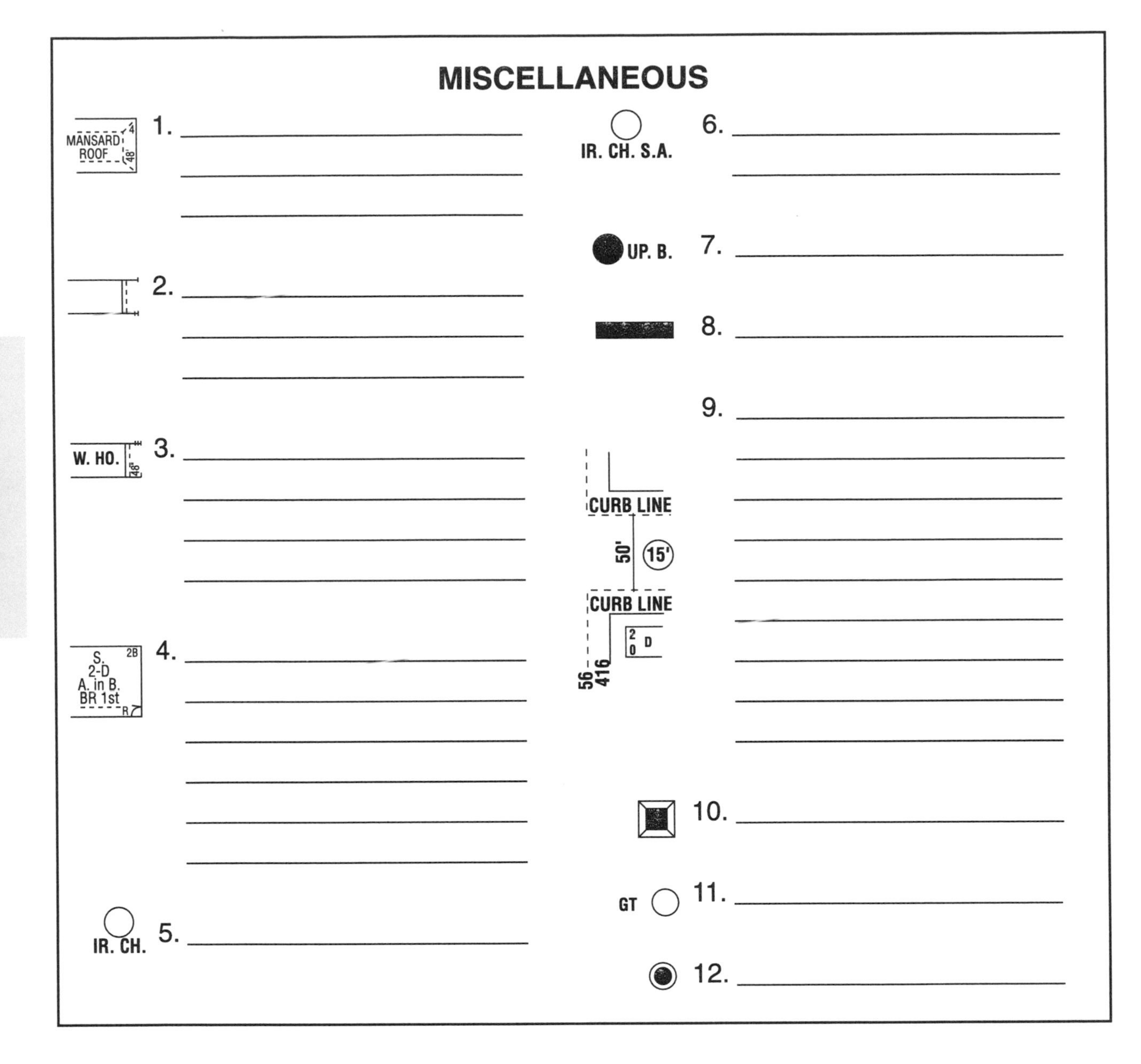
MISCELLANEOUS
MANSARD ROOF
4
48"
1.
2.
W. HO.
48"
3.
S.
2B
2-D
A. in B.
BR 1st
R
4.
IR. CH.
5.
IR. CH. S.A.
6.
UP. B.
7.
8.
9.
CURB LINE
50'
15'
CURB LINE
2
0
D
56
416
10.
GT
11.
12.

COLOR CODE FOR CONSTRUCTION

Materials for Walls

Brown 1. ______________________________

Red 2. ______________________________

Yellow 3. ______________________________

Blue 4. ______________________________

Gray 5. ______________________________

Answers

Chapter 1 Answers

FIREFIGHTER I

Matching

A.

1. g *(5)*
2. d *(7)*
3. a *(7)*
4. f *(7)*
5. j *(7)*
6. c *(7)*
7. i *(12)*
8. l *(12)*
9. e *(12)*
10. k *(12)*
11. h *(19)*

True/False

B.

1. False. New firefighters enter into one of *three categories: career, paid on call, or volunteer. (5)*
2. False. Whenever there is an emergency, the fire department is one of the first entities called to the scene. *The emergency list is unlimited. (5)*
3. False. Firefighters are not extraordinary — *they are ordinary people. They cannot do everything at once. (5)*
4. True *(8, 9)*
5. True *(10)*
6. False. The special rescue technician *handles special rescue situations such as high-angle rescue, trench and structural collapse, confined space entry, etc. (11)*
7. True *(11)*
8. True *(11)*
9. False. *A firefighter's training never ends. (12)*
10. False. To ensure that department members cooperate effectively, *the methods of doing so are outlined in policies and procedures. (12)*
11. True *(12)*
12. False. *Procedures have a built-in flexibility* that allows, with reasonable justification, *adjustments when unforeseen circumstances occur. (13)*
13. True *(13)*
14. True *(13)*
15. True *(13)*
16. False. SOPs may include *regulations on dress, conduct, vacation and sick leave, station life and duties, and other departmental policies. (14)*
17. True *(19)*
18. True *(20)*
19. False. Most firefighter injuries are a direct result of *preventable accidents. (21)*
20. False. The program should be readily available *to all members and their families. (22)*
21. True *(23)*
22. True *(23)*
23. False. *Do not ride on the tailboard.* Many firefighters have been killed falling from tailboards. *(23)*
24. True *(25)*
25. False. The use of a cheater can overload a tool, not only while the cheater is attached, *but also later when the weakened tool is being used normally. (26)*
26. False. If trainees have colds, severe headaches, or other symptoms indicating physical discomfort or illness, *they should not continue training. (28)*
27. True *(28)*
28. False. The IC decides when the *risks involved are great enough to warrant limiting the actions* of the fire fighting personnel. *(28)*
29. True *(29)*

Multiple Choice

C.

1. d *(5)*
2. b *(5)*
3. a *(7)*
4. d *(7)*
5. c *(7)*
6. d *(8)*

FIREFIGHTER I

7. d *(9)*
8. c *(10)*
9. a *(10)*
10. a *(11)*
11. b *(11)*
12. d *(12)*
13. a *(13)*
14. b *(13, 14)*
15. a *(19)*
16. c *(19)*
17. d *(20)*
18. d *(21)*
19. b *(22)*
20. b *(22)*
21. d *(22)*
22. a *(23)*
23. b *(23, 24)*
24. c *(25)*
25. b *(25)*
26. a *(26)*
27. d *(26)*
28. c *(26)*
29. d *(26, 27)*
30. a *(27, 28)*
31. b *(28)*
32. c *(28)*
33. d *(28)*
34. c *(29)*
35. d *(29)*
36. a *(30)*

Identify

D.

1. National Fire Protection Association *(8)*
2. Aircraft rescue and fire fighting *(10)*
3. Nuclear, biological, and chemical *(10)*
4. Self-contained underwater breathing apparatus *(10)*
5. Emergency medical services *(11)*
6. Emergency medical technician *(11)*
7. Basic life support *(11)*
8. Advanced life support *(12)*
9. Standard operating procedure *(12)*
10. Self-contained breathing apparatus *(13)*
11. Intravenous *(19)*
12. Environmental Protection Agency *(20)*
13. Employee assistance program *(22)*
14. Critical incident stress debriefing *(23)*
15. Personal protective equipment *(26)*

FIREFIGHTER II

Matching

A.

1. e *(15)*
2. a *(15)*
3. d *(15)*
4. b *(14)*

B.

1. d *(15)*
2. b *(15)*
3. e *(16)*
4. h *(16)*
5. g *(16)*
6. f *(16)*
7. c *(16)*

True/False

C.

1. False. It is designed to be applicable to *incidents of all sizes and types. (14)*
2. True *(14)*
3. False. Operations may be subdivided into as many as *five branches* if necessary. *(15)*
4. False. Generally, Finance will be activated *only on large-scale, long-term incidents. (15)*
5. True *(15)*
6. False. Small, routine incidents *usually do not require a written plan. (16)*
7. True *(16)*
8. False. With advice from the Operations Officer, *the IC* will gather resources to handle the incident. *(17)*
9. False. *All incident personnel* must function according to the Incident Action Plan. *(17)*
10. True *(17)*
11. False. Command can only be transferred to *someone on the scene. (17)*
12. True *(18)*
13. False. One of the *most important functions* of the IMS *is to provide a means of tracking all personnel and equipment assigned to the incident. (18)*
14. False. Once the incident has been brought under control and the size and complexity of the situation diminishes, *the resources that are no longer needed should be released. (18)*

Multiple Choice

D.

1. d *(14)*
2. b *(14, 15)*
3. c *(15)*
4. d *(15)*
5. c *(15)*
6. a *(16)*
7. a *(16)*
8. c *(16)*
9. c *(16)*
10. d *(16, 17)*
11. a *(17)*
12. d *(17)*
13. a *(17)*
14. b *(17)*
15. b *(18)*
16. d *(18)*
17. c *(18)*
18. a *(18)*

Identify

E.

1. Incident Management System *(14)*
2. Incident Commander *(16)*
3. Incident Action Plan *(16)*
4. Command Post *(17)*

Chapter 2 Answers

FIREFIGHTER I

Matching

A.

1. g *(33)*
2. a *(33)*
3. d *(33)*
4. j *(42)*
5. b *(42)*
6. h *(42)*
7. e *(43)*
8. k *(45)*
9. i *(47)*
10. f *(53)*

B.

1. c *(34)*
2. f *(34)*
3. a *(34)*
4. i *(35)*
5. e *(35)*
6. k *(35, 36)*
7. j *(36)*
8. b *(36)*
9. d *(36)*
10. g *(38)*

C.

1. f *(34)*
2. b *(34)*
3. d *(35)*
4. a *(36)*
5. b *(36)*
6. g *(36)*
7. c *(39)*

True/False

D.

1. False. Many of the concepts hold true for wildland fires, but *a number of additional factors must be addressed in those incidents. (33)*
2. True *(34)*
3. False. *SI* is a very logical and simple system based on powers of 10. *(34)*
4. True *(37)*
5. False. Good insulators are materials that do not conduct heat well because their *physical makeup disrupts the point-to-point transfer of heat energy. (37)*
6. True *(38)*
7. False. The other factor is pressure. *As the pressure on the surface of a substance decreases, so does the temperature at which it boils. (39)*
8. True *(39)*
9. True *(39)*
10. False. The oxidation reaction *releases energy or is exothermic. (40)*
11. False. Because oxidation is an *exothermic* process, *it always produces heat. (40)*
12. False. While the fire triangle is useful, *it is not technically correct.* For combustion, four components are necessary. *(40)*
13. False. While oxygen is the most common oxidizer, *other substances also fall into this category. (41)*
14. True *(42)*
15. True *(42)*
16. True *(43)*
17. False. When contained, the specific volume of a liquid has a relatively *low* surface-to-volume ratio. *(43)*
18. False. Materials with different heats of combustion are converted to be equivalent to *the heat of combustion of wood. (45)*
19. True *(45)*
20. False. Ignition describes the period when *the four elements of the fire tetrahedron* come together and combustion begins. *(48)*
21. False. While *no exact temperature* is associated with this occurrence, a range from approximately 900°F to 1,200°F *(483°C to 649°C)* is widely used. *(50)*
22. True *(51)*
23. False. Firefighters who find themselves in a compartment at flashover are at extreme risk *even while wearing their personal protective equipment. (51)*

24. True *(51, 52)*
25. False. Firefighters should be able *to recognize potential fuel packages* in a building or compartment and *use this information to estimate the fire growth potential. (52)*
26. True *(53)*
27. True *(54, 55)*
28. False. *Smoke* causes most deaths in fires. *(55)*
29. True *(56)*
30. False. If water is used, the *fuel can float on it while continuing to burn.* If the fuel is unconfined, *using water could unintentionally spread a fire. (58)*
31. True *(58)*
32. False. Gases or vapors with vapor densities *greater* than 1 *tend to hug the ground and travel as directed by terrain and wind. (58)*
33. True *(58)*
34. False. CO_2 flooding or coating with foam *does not provide the cooling effect* needed for total extinguishment. *(58)*
35. True *(58)*
36. True *(60)*
37. False. Information regarding a material and its characteristics should be reviewed *prior to attempting to extinguish a Class D fire. (60)*

Multiple Choice

E.

1. b *(34)*
2. a *(35)*
3. c *(36)*
4. a *(35)*
5. d *(36)*
6. b *(36)*
7. d *(36)*
8. a *(36, 37)*
9. b *(37)*
10. c *(37, 38)*
11. c *(38)*
12. a *(39)*
13. d *(39)*
14. a *(39)*
15. c *(39)*
16. b *(40)*
17. a *(40)*
18. c *(40)*
19. d *(40)*
20. a *(42)*
21. b *(42)*
22. b *(42)*
23. c *(42)*
24. a *(42, 43)*
25. c *(43)*
26. c *(44)*
27. b *(44)*
28. d *(45)*
29. b *(45)*
30. c *(45)*
31. c *(46)*
32. a *(47)*
33. b *(47)*
34. d *(48)*
35. c *(48)*
36. a *(49)*
37. a *(49)*
38. b *(52)*
39. c *(52)*
40. d *(53)*
41. b *(53, 54)*
42. c *(55)*
43. d *(56)*
44. c *(57)*
45. a *(56)*

Identify

F.

1. International System of Units *(34)*
2. National Institute of Standards and Technology *(35)*
3. Heat release rate *(35)*
4. Lower flammable limit *(44)*
5. Upper flammable limit *(44)*
6. Carbon monoxide *(56)*
7. Hydrogen cyanide *(56)*
8. Carbon dioxide *(56)*
9. Material Safety Data Sheet *(60)*
10. *North American Emergency Response Guidebook (60)*

G.

1. Plume development *(50)*
2. Flashover *(51)*
3. Rollover *(53)*
4. Backdraft *(55)*

H. *(57)*

1. C
2. A
3. D
4. B

I.

1. Class D *(60)*
2. Class B *(59)*
3. Class A *(59)*
4. Class C *(59)*

Chapter 3 Answers

FIREFIGHTER I

Matching

A.

1. d *(69)*
2. a *(71)*
3. e *(72)*
4. b *(72)*
5. c *(75)*

True/False

B.

1. False. *All firefighters* should have a basic knowledge of the principles of building construction. *(65)*
2. True *(65)*
3. False. Materials with no fire-resistance ratings *may be used in limited quantities. (66)*
4. True *(66)*
5. True *(67)*
6. False. This type of construction presents almost unlimited potential for fire extension to nearby structures. Firefighters *must be alert for fire coming from doors or windows. (67)*
7. False. Fire retardants are *not always effective* in reducing fire spread. *(68)*
8. False. Water used during extinguishing operations does *not* have a substantial negative effect on the structural strength of wood construction materials. *(68)*
9. True *(69, 70)*
10. True *(70)*
11. False. Reinforced concrete does *not perform particularly well* under fire conditions. *(70)*
12. False. The water content gives gypsum *excellent heat-resistant, fire-retardant properties. (71)*
13. False. Wood shake shingles in particular, *even when treated with fire retardant,* can *significantly contribute to fire spread. (72)*
14. True *(73)*
15. False. *No personnel or apparatus* should be allowed to operate in the collapse zone except to place unmanned master stream devices. *(74)*
16. False. Although the trusses may be protected with fire-retardant treatments to give longer protection, *most are not protected at all. (75)*
17. True *(76)*
18. True *(76)*

Multiple Choice

C.

1. a *(65)*
2. d *(66)*
3. d *(66)*
4. a *(66)*
5. b *(66)*
6. d *(67)*
7. c *(67)*
8. b *(68)*
9. c *(68)*
10. b *(68)*
11. b *(69)*
12. d *(69)*
13. a *(69)*
14. a *(70)*
15. a *(70)*
16. b *(70)*
17. b *(70)*
18. d *(70, 71)*
19. c *(71)*
20. b *(72)*
21. a *(72)*
22. a *(72)*
23. b *(73)*
24. b *(73)*
25. a *(74)*
26. b *(74)*
27. a *(75)*
28. d *(76)*
29. c *(76)*
30. c *(75)*

Identify

D. *(65)*

1. Type II
2. Type V
3. Type I
4. Type III
5. Type IV

FIREFIGHTER II

List

A. *(73, 74)*

1. Cracks or separations in walls, floors, ceilings, and roof structures
2. Evidence of existing structural instability (e.g., tie rods and stars that hold walls together)
3. Loose bricks, blocks, or stones falling from buildings
4. Deteriorated mortar between the masonry
5. Walls that appear to be leaning
6. Structural members that appear to be distorted
7. Fires beneath floors that support heavy machinery or other extreme weight loads
8. Prolonged fire exposure to the structural members
9. Unusual creaks and cracking noises
10. Structural members pulling away from walls
11. Excessive weight of building contents

B. *(74)*

1. Improper vertical ventilation techniques can result in the cutting of structural supports that could weaken the structure.
2. The water used to extinguish a fire adds extra weight to the structure and can weaken it.

Chapter 4 Answers

FIREFIGHTER I

Matching

A.

1. c *(85)*
2. e *(88)*
3. a *(88)*
4. d *(90)*
5. g *(95)*
6. f *(108)*

B. *(89, 90)*

1. d
2. g
3. a
4. e
5. b
6. c

C. *(89, 90)*

1. e
2. a
3. b
4. c
5. d
6. g

D. *(89, 90)*

1. a
2. b
3. e
4. c
5. g
6. d

True/False

E.

1. True *(79)*
2. False. *All firefighters operating at an emergency scene* must wear full protective equipment. *(79)*
3. False. Each component of the protective ensemble must have an appropriate product label for that component *permanently and conspicuously attached. (80)*
4. True *(81)*
5. True *(81)*
6. False. Care must be taken to ensure that the hood *does not interfere with the facepiece-to-face seal. (82)*
7. False. Three-quarter boots and long coats *alone do not provide adequate protection. (83)*
8. True *(84)*
9. False. Wear *protective boots during fire fighting and emergency activities* and *safety shoes for station wear* and other fire department activities. *(84)*
10. False. Firefighters should *not* share protective boots because *this practice is unsanitary. (84, 85)*
11. True *(85)*
12. False. While this clothing is designed to be fire resistant, it is *not designed to be worn for fire fighting operations. (86)*
13. True *(87)*
14. False. The tissue damage from inhaling hot air *is not immediately reversible* by introducing fresh, cool air. *(88)*
15. False. Black smoke is *high in particulate carbon and carbon monoxide* because of incomplete combustion. *(90)*
16. True *(90)*
17. True *(90, 91)*
18. True *(91)*
19. False. Hazardous atmospheres can be found in *numerous situations in which fire is not involved. (91)*
20. False. The atmosphere in many of these areas is *oxygen deficient* and *will not support life* even though there may be no toxic gas. *(92)*
21. True *(93)*
22. True *(95)*
23. False. Such a substitution *voids NIOSH and MSHA certification* and is not a recommended practice. *(95)*
24. False. The audible alarm sounds when the cylinder pressure decreases to approximately *one-fourth* of the maximum rated pressure. *(97)*
25. True *(99)*

26. False. Do *not* keep the facepiece connected to the regulator during storage. *These parts must be separate* to check for proper facepiece seal. *(102)*
27. False. NFPA 1500 requires firefighters *to remain seated and belted at all times* while the emergency vehicle is in motion. *(102)*
28. True *(105)*
29. False. After checking the bypass valve, *make sure it is fully closed. (105)*
30. False. All firefighters should be aware of the signs and symptoms of heat-related conditions. *Know your own limitations and abilities. (107)*
31. True *(107)*
32. False. The firefighter should *never remove the SCBA facepiece to negotiate a restricted area. (109, 110)*

Multiple Choice

F.

1. a *(80)*
2. b *(80, 81)*
3. c *(81)*
4. d *(81)*
5. b *(81)*
6. d *(82)*
7. a *(81)*
8. c *(82)*
9. a *(82)*
10. d *(82)*
11. b *(83)*
12. d *(83)*
13. c *(83)*
14. b *(83, 84)*
15. a *(84)*
16. d *(85)*
17. d *(85)*
18. d *(85)*
19. a *(85)*
20. c *(85, 86)*
21. a *(86)*
22. c *(86)*
23. a *(86)*
24. b *(87)*
25. c *(87)*
26. c *(87)*
27. b *(87)*
28. a *(88)*
29. d *(89)*
30. b *(89)*
31. b *(88)*
32. c *(91)*
33. c *(89)*
34. a *(89)*
35. b *(89)*
36. a *(90)*
37. c *(89)*
38. c *(93)*
39. a *(90)*
40. d *(90)*
41. b *(90)*
42. c *(90)*
43. d *(90)*
44. a *(91)*
45. a *(93)*
46. b *(95, 96)*
47. a *(96)*
48. d *(96)*
49. c *(96, 97)*
50. a *(97)*
51. d *(97)*
52. c *(98)*
53. a *(99)*
54. b *(100)*
55. b *(100)*
56. c *(100)*
57. b *(101)*
58. d *(102)*
59. c *(103, 104)*
60. a *(104)*
61. d *(105)*
62. a *(105)*
63. a *(105)*
64. d *(106)*
65. a *(106)*
66. c *(106)*
67. b *(106, 107)*
68. a *(108)*
69. d *(108)*
70. d *(109)*
71. c *(113)*

Identify

G.

1. Self-contained breathing apparatus *(80)*
2. Personal alert safety system *(80)*
3. Immediately dangerous to life and health *(89)*
4. National Institute for Occupational Safety and Health *(89)*
5. Carbon monoxide *(89)*
6. Carboxyhemoglobin *(90)*
7. Part per million *(90)*
8. Department of Transportation *(93)*
9. Occupational Safety and Health Administration *(93)*
10. Mine Safety and Health Administration *(95)*
11. Personal alert device *(99)*

H. *(104)*

a. 5
b. 2
c. 1
d. 4
e. 7
f. 6
g. 3

FIREFIGHTER I

Chapter 5 Answers

FIREFIGHTER I

Matching

A.

1. c *(127)*
2. g *(127)*
3. a *(128)*
4. d *(129)*
5. h *(130)*
6. e *(129, 130)*
7. f *(130)*

True/False

B.

1. True *(125)*
2. False. NFPA 1901 requires that pumping apparatus have *three approved portable fire extinguishers and mounting brackets. (125)*
3. True *(128)*
4. False. Freeze protection may be provided by *adding antifreeze to the water* or by storage in a warm area. *(127)*
5. False. The foam is *ineffective on flammable liquids that are water-soluble such as alcohol and acetone. (128)*
6. True *(128)*
7. False. Halogenated vapor is nonconductive and *is effective* in extinguishing surface fires in flammable and combustible liquids and electrical equipment. *(129)*
8. True *(129)*
9. True *(130)*
10. False. Carbon dioxide produces *no vapor-suppressing film;* therefore, the *reignition of the fuel is always a danger. (130)*
11. False. The terms *dry chemical* and *dry powder* are often incorrectly used interchangeably. *Dry powder agents are for Class D fires only. (130)*
12. True *(131)*
13. False. Dry chemicals agents themselves *are nontoxic and generally considered quite safe to use* but may reduce visibility and create respiratory problems like any airborne particulate. *(131)*
14. True *(132)*
15. True *(132)*
16. False. Extinguishers suitable for more than one class of fire are identified by *combinations of the letters A, B, and/or C or the symbols for each class. (134)*
17. True *(134)*
18. False. Select extinguishers that *minimize the risk to life and property* but are effective in extinguishing the fire. *(136)*
19. False. Firefighters *should become familiar with the detailed instructions* found on the label of the extinguisher. *(136)*
20. False. Lay empty fire extinguishers *on their sides after use. (137)*
21. True *(137)*
22. True *(139)*

Multiple Choice

C.

1. d *(125)*
2. a *(125)*
3. a *(125, 127)*
4. b *(127)*
5. a *(127)*
6. c *(128)*
7. d *(128)*
8. b *(129)*
9. c *(125, 127)*
10. a *(130)*
11. d *(130)*
12. b *(131, 132)*
13. a *(133)*
14. b *(134)*
15. c *(134, 135)*
16. a *(136)*
17. a *(137)*
18. d *(137, 138)*
19. a *(138)*
20. c *(139)*

Identify

D.

1. Carbon dioxide *(125)*
2. Air-pressurized water *(127)*
3. Aqueous film forming foam *(128)*
4. Halogenated hydrocarbons *(129)*
5. Underwriters Laboratories Inc. *(134)*
6. Underwriters Laboratories of Canada *(134)*

E. *(138)*

1, 2, 4, 6

F. *(135)*

1. Class D
2. Class C
3. Class B
4. Class A

G. *(135)*

1. Class A
2. Class B
3. Class C

H.

1. Hand-carried halon *(129)*
2. Handheld carbon dioxide *(130)*
3. AFFF *(128)*
4. Dry chemical (cartridge-operated) *(131)*
5. Stored-pressure water *(127)*
6. Dry chemical (stored-pressure) *(131)*
7. Pump-tank water *(127)*
8. Handheld Class D *(132)*
9. Wheeled halon *(129)*
10. Wheeled carbon dioxide *(130)*
11. Wheeled dry chemical *(131)*
12. Wheeled Class D *(132)*

Chapter 6 Answers

FIREFIGHTER I

Matching

A.

1. d *(148, 150)*
2. a *(150)*
3. g *(154)*
4. f *(154)*
5. c *(154)*
6. b *(155)*

B.

1. e *(156)*
2. g *(156)*
3. f *(156)*
4. b *(156)*
5. a *(155)*
6. d *(155, 156)*
7. c *(155)*

True/False

C.

1. False. Life safety rope is used during actual incidents *or training.* Utility rope is used in any instance, *excluding life safety applications. (147)*
2. True *(148)*
3. False. Any life safety rope that has been impact-loaded *should be destroyed immediately. (148)*
4. False. Synthetic fiber rope has *excellent resistance to mildew and rotting. (148)*
5. True *(150)*
6. False. *Half of its strength is in the sheath* and the other half is in the core. *(150)*
7. True *(151)*
8. False. The presence of mildew *does not necessarily indicate a problem;* however, the rope should be cleaned and reinspected. *(151)*
9. False. *When a piece of rescue rope is purchased,* it should be permanently identified and a record (rope logbook) started and kept throughout its working life. *(152)*
10. True *(153)*
11. False. These devices do an adequate job of cleaning mud and other surface debris, *but for a more thorough cleaning, rope should be washed in a clothes-washing machine. (153)*
12. True *(154)*
13. True *(154)*
14. False. A rope's strength is reduced whenever it is bent. *The tighter the bend, the more strength is lost. (154)*
15. False. It is *unlikely to slip* when the rope is wet. *(156)*
16. True *(156)*
17. False. Hoisting hose is *possibly the safest way* of getting hoselines to upper levels. *(157)*
18. False. Rope rescue is a *technical skill that requires specialized training. (158)*

Multiple Choice

D.

1. c *(147)*
2. b *(148)*
3. d *(148)*
4. d *(148)*
5. a *(149)*
6. b *(150)*
7. d *(150)*
8. b *(150)*
9. a *(150)*
10. d *(150)*
11. c *(150)*
12. b *(150)*
13. a *(150)*
14. d *(150)*
15. d *(150)*
16. a *(151)*
17. b *(152)*
18. c *(153)*
19. c *(153)*
20. d *(153)*
21. d *(153)*
22. a *(153)*
23. c *(153, 154)*
24. c *(167)*
25. d *(164)*
26. a *(163)*
27. a *(168)*
28. d *(162)*
29. a *(155)*
30. a *(154)*
31. c *(155)*
32. b *(155, 156)*
33. c *(156)*
34. c *(156)*
35. a *(156)*
36. b *(156)*
37. d *(157)*
38. a *(157)*
39. c *(157)*

Identify

E. *(149)*

1. 3
2. 100%
3. 3
4. 500°F *(260°C)*
5. Good
6. 100%
7. 15–20%
8. Wet or dry
9. Floats (0.95)
10. 6

F. *(149)*

1. 5
2. Poor
3. 115%
4. 5–10%
5. Good
6. Dry only
7. 2
8. 5
9. Floats (0.97)
10. 275°F *(135°C)*
11. Excellent

Chapter 7 Answers

FIREFIGHTER I

Matching

A.

1. c *(175)*
2. e *(175)*
3. b *(176)*
4. d *(185, dictionary)*
5. a *(185, dictionary)*
6. f *(176, 178)*

True/False

B.

1. False. *Regardless of how small a structure fire may look upon arrival,* the fire department must *always* do a thorough search of the building. *(175)*
2. False. Firefighters should not assume that all occupants are out until the building has been *searched by firefighters. (175)*
3. False. During the primary search, rescuers should *always use the buddy system. (176)*
4. False. Primary search personnel should *always carry forcible entry tools* with them whenever they enter a building and throughout the search. *(176)*
5. True *(178)*
6. False. During the secondary search, *speed is not as critical as thoroughness. (178)*
7. True *(178)*
8. False. While searching for victims in a fire, *rescuers must always consider their own safety. (180)*
9. True *(181)*
10. False. *Even* with the best incident command or accountability system in place, *unusual circumstances can lead to a firefighter becoming trapped or disoriented within a burning structure. (181)*
11. False. These firefighters should *immediately activate their PASS devices* and try to maintain composure to maximize their air supplies. *(182)*
12. False. At *no time should rescuers remove their facepieces* or in any way compromise the proper operation of their SCBA to share them with another firefighter or victim. *(182)*
13. True *(183)*
14. False. Firefighters should work from a single action plan. *Crews should not be allowed to freelance. (183)*
15. True *(206)*

Multiple Choice

C.

1. c *(175)*
2. b *(175)*
3. a *(176)*
4. b *(176)*
5. d *(176)*
6. c *(177)*
7. c *(177)*
8. a *(177)*
9. d *(177)*
10. a *(178)*
11. b *(179)*
12. b *(179)*
13. c *(179)*
14. a *(179, 180)*
15. c *(181)*
16. a *(181)*
17. d *(181, 182)*
18. d *(182)*
19. a *(183)*
20. c *(183, 184)*
21. b *(184)*
22. d *(185)*
23. c *(185)*
24. a *(185)*
25. b *(185)*
26. b *(206)*

Identify

D.

1. Incident commander *(176)*
2. Cardiopulmonary resuscitation *(184)*

FIREFIGHTER II

Matching

A.

1. e *(186)*
2. d *(190, 192, dictionary)*
3. c *(202)*
4. j *(202)*
5. i *(208)*
6. a *(208)*
7. f *(211)*
8. b *(196)*
9. g *(196)*

True/False

B.

1. False. *Generators* are the most common power source used for emergency services. *(186)*
2. True *(186)*
3. False. Mutual aid departments who frequently work together and have either different sizes or different types of receptacles *should carry adapters so that equipment can be interchanged. (188)*
4. True *(189)*
5. False. The combination tool's spreading and cutting capabilities *are somewhat less than those of the individual units. (189)*
6. True *(191)*
7. False. Rescuers should *never work under a load supported only by a jack. (191)*
8. True *(192)*
9. True *(193)*
10. False. The bags should be positioned *on or against a solid surface. (196)*
11. False. They should be used *only to lift or stabilize objects and NOT people. (197)*
12. True *(197)*
13. False. While each vehicle is being checked, *another rescuer should be assigned* to survey the entire area around the scene. *(197, 198)*
14. False. Rescuers must be trained to *resist testing the vehicle's stability* because the slightest push in the wrong way could cause the vehicle to move. *(198)*
15. True *(200)*
16. True *(201)*
17. True *(202)*
18. False. If it is necessary to break a window to gain primary access to a victim, choose one *as far away from the victim as possible. (204)*
19. False. Plastics used in vehicle construction *may prevent the roof from bending,* and unibody vehicles have *features that affect them when their roofs are removed. (205)*
20. True *(210, 211)*
21. False. Just because ice is thick *does not mean that it is strong. (212)*
22. False. *Under no circumstances* should firefighters alter the elevator's mechanical system in an attempt to move the elevator. *(214)*
23. False. Communication with the passengers *is essential for their morale and mental state and should be established and maintained throughout the operation. (214)*
24. True *(214)*
25. False. Shoring *is not intended to move heavy objects* but is just intended to stabilize them. *(208)*

Multiple Choice

C.

1. d *(186)*
2. b *(187)*
3. a *(187)*
4. a *(188)*
5. b *(188)*
6. b *(189)*
7. c *(190)*
8. a *(190)*
9. c *(191)*
10. d *(191)*
11. d *(192)*
12. d *(192)*
13. c *(193)*
14. b *(193)*
15. a *(194)*
16. d *(194)*
17. a *(196)*
18. c *(196)*
19. b *(206, 207)*
20. a *(207)*
21. d *(207)*
22. c *(207)*

23. a *(207, 208)*
24. d *(197)*
25. c *(197)*
26. a *(199)*
27. b *(201)*
28. d *(201)*
29. c *(203)*
30. d *(204)*
31. c *(205)*
32. d *(208)*
33. a *(208)*
34. c *(209)*
35. b *(210)*
36. a *(210)*
37. c *(212)*
38. b *(212, 213)*
39. a *(213)*
40. a *(213)*

Identify

D.

1. Supplemental Restraint System *(201)*
2. Side-Impact Protection System *(201)*
3. Personal flotation device *(211)*

E.

1. Bar screw jack *(191)*
2. Combination spreader/shears *(189)*
3. Ratchet-lever jack *(191)*
4. Hydraulic shears *(189)*
5. Hydraulic spreaders *(189)*
6. Hydraulic jack *(190)*
7. Extension rams *(189)*
8. Trench screw jack *(191)*

Chapter 8 Answers

FIREFIGHTER I

Matching

A.

1. f *(233)*
2. d *(271, dictionary)*
3. a *(255, dictionary)*
4. e *(256, dictionary)*
5. g *(240)*
6. c *(264)*

True/False

B.

1. False. *Forcible entry is a learned skill.* It is not easy and *must be practiced often. (233)*
2. True *(233)*
3. False. There is *no such thing as a single cutting tool that will efficiently cut all materials. (234)*
4. False. Handsaws are *extremely slow. (234)*
5. False. *Never use a power saw in a flammable atmosphere. (235)*
6. True *(236)*
7. False. *Some* prying tools can be used effectively as striking tools, although *most cannot. (237)*
8. True *(240)*
9. True *(240)*
10. False. When entering a building, *carefully invert the tool and carry it with the head upright close to the body. (241)*
11. False. *Never carry a power tool that is running.* Transport the tool to the area where the work will be performed and start it there. *(241)*
12. False. *The core or center portion of the door is made up of a web or grid of glued wooden strips* over which several layers of veneer panels have been glued. *(244)*
13. True *(245)*
14. False. *It is more effective to force through the swinging door* than trying to force open a locked revolving door. *(246)*
15. False. *All overhead doors should be blocked open* to prevent injury to firefighters should a control device fail. *(249)*
16. True *(250)*
17. True *(250)*
18. False. If breaking glass to gain access into a fire building, *SCBA should be worn* and a charged hoseline should be in place, ready to attack the fire. *(253)*
19. False. If the door and lock are *suitable for conventional forcible entry, then it should be used. (255)*
20. False. *Do not try to cut a loose padlock alone.* Work with a partner. *(258)*
21. False. Forcible entry can take place through windows, *though they are not the preferred entry point into a fire building. (260)*
22. False. *Lexan® is 250 times stronger than safety glass. (263)*
23. True *(265)*
24. True *(266)*
25. False. *A wood floor does not in itself ensure that it can be penetrated easily.* Many wood floors are laid over a concrete slab. *(266)*
26. False. It is *better to supply power to electric saws from a portable generator* carried on the fire apparatus than to depend on domestic power during a fire. *(266)*
27. True *(266)*

Multiple Choice

C.

1. d *(234)*
2. b *(234, 235)*
3. c *(235)*
4. a *(235)*
5. b *(236)*
6. d *(236)*

7. a *(237)*
8. c *(237)*
9. a *(238)*
10. b *(238)*
11. d *(238, 239)*
12. d *(239)*
13. a *(240)*
14. c *(240)*
15. b *(240)*
16. d *(241)*
17. a *(242)*
18. a *(242)*
19. c *(242, 243)*
20. d *(243)*
21. d *(243)*
22. b *(243)*
23. a *(244)*
24. c *(244)*
25. d *(244)*
26. b *(244, 245)*
27. a *(245)*
28. c *(245)*
29. c *(245)*
30. a *(246)*
31. d *(247)*
32. b *(247)*
33. c *(249)*
34. c *(249)*
35. b *(250)*
36. d *(251)*
37. c *(250)*
38. b *(252)*
39. c *(251)*
40. b *(252)*
41. a *(252)*
42. b *(254)*
43. c *(254)*
44. d *(254)*
45. d *(253, 255)*
46. b *(255)*
47. a *(256)*
48. a *(256, 257)*
49. c *(257)*
50. b *(258)*
51. d *(258, 259)*
52. c *(259)*
53. a *(260)*
54. d *(261)*
55. a *(261)*
56. b *(261)*
57. c *(261)*
58. d *(262)*
59. a *(262)*
60. b *(263)*
61. d *(265)*
62. b *(265)*
63. c *(266)*
64. c *(266)*
65. b *(266)*
66. c *(266, 267)*

Identify

D.

1. Bolt cutters *(236)*
2. Hacksaw *(234)*
3. Flat-head axe *(239)*
4. Rotary (circular saw) *(235)*
5. Cutting torch *(236)*
6. Keyhole saw *(234)*
7. Coping saw *(234)*
8. Pick-head axe *(234)*
9. Chain saws *(235)*
10. Reciprocating saw *(235)*
11. Carpenter's saw *(234)*

E.

1. Roofman's hook *(238)*
2. Clemens hook *(238)*
3. Multipurpose hook *(238)*
4. Plaster hook *(238)*
5. Pike pole *(238)*
6. San Francisco hook *(238)*
7. Drywall hook *(238)*

F.

1. Claw tool *(237)*
2. Flat bar *(237)*
3. Crowbar *(237)*
4. Halligan bar *(237)*
5. Pry bar *(237)*
6. Kelly tool *(237)*
7. Hux bar *(237)*

G. *(239)*

1. Sledgehammer
2. Battering ram
3. Maul
4. Punches
5. Pick
6. Mallet
7. Ball peen hammer
8. Claw hammer

H.

1. Frame-and-brace door *(245)*
2. Frame-and-ledge door *(245)*
3. Ledge door *(245)*
4. Hollow core door *(244)*
5. Solid core door *(244)*
6. Stopped jamb *(244)*
7. Rabbeted jamb *(244)*

I.

1. Slab-type garage door *(248)*
2. Pocket door *(246)*
3. Rolling steel door *(249)*
4. Revolving door *(246)*

J.

1. Key-in-knob lock *(251)*
2. Rim lock (interlocking dead bolt) *(252)*
3. Padlocks *(252)*
4. Mortise lock *(251)*

K.

1. A-tool *(256)*
2. Shove knife *(257)*
3. K-tool *(256)*
4. J-tool *(257)*

L. *(257)*

1. Bam-bam tool
2. Locking pliers and chain
3. Duck-billed lock breaker
4. Hockey puck lock breaker
5. Hammerheaded pick

M.

1. Jalousie louvered *(263)*
2. Projected (factory) *(262)*
3. Casement (hinged) *(261)*
4. Awning louvered *(262)*
5. Double-hung (checkrail) *(260)*

Chapter 9 Answers

FIREFIGHTER I

Matching

A.

1. c *(281)*
2. h *(281)*
3. o *(282)*
4. a *(282)*
5. l *(282)*
6. f *(282)*
7. d *(283)*
8. b *(283)*
9. j *(283)*
10. e *(283)*
11. m *(284)*
12. i *(284)*
13. g *(284)*
14. k *(284)*

B.

1. e *(285)*
2. b *(285)*
3. g *(286)*
4. a *(286)*
5. f *(287)*
6. d *(287)*

True/False

C.

1. True *(286)*
2. False. *All* ladders meeting NFPA 1931 are *required to have a certification label affixed to the ladder by the manufacturer* indicating that the ladder meets the standard. *(287)*
3. False. *Any* indication of deterioration of wood *is cause for the ladder to be removed from service* until it can be service tested. *(289)*
4. False. Ladders that cannot be safely repaired *have to be destroyed or scrapped for parts. (290)*
5. True *(290)*
6. False. Contact with power sources *may result in electrocution of anyone in contact with the ladder. (291)*
7. False. The designated length *is NOT THE LADDER'S REACH.* Ladders are set at angles of approximately 75 degrees; therefore, the reach will be LESS than the designated length. *(291, 292)*
8. False. There are *no established standards* for the location and mounting of ground ladders on fire apparatus. *(292)*
9. True *(293)*
10. True *(297)*
11. False. If the ladder is too close to the building, *its stability is reduced and it tends to cause the tip to pull away from the building.* If the butt is too *far* from the building, *the load-carrying capacity is reduced. (298)*
12. False. *Only the butt* needs to be placed on the ground. *(299)*
13. False. Many firefighters do not realize that WET wood and *fiberglass ladders present electrocution hazards. (299)*
14. False. This distance must be maintained at all times, *including during the raise itself. (299)*
15. True *(301)*
16. False. When a roof ladder has been carried to the scene butt first, *there is no need to waste valuable time turning it around.* Firefighters should use the butt-first placement method. *(302)*
17. True *(303)*
18. False. If the ladder is properly spaced from the building, *the firefighter's body will be perpendicular to the ground. (304)*
19. True *(304)*
20. False. The climber's *arms should be kept straight* during the climb; *this action keeps the body away from the ladder* and permits free knee movement during the climb. *(304)*
21. False. Practice climbing should *be done slowly to develop form* rather than speed. *(304)*
22. True *(305)*
23. True *(305)*

Multiple Choice

D.

1. b *(281)*
2. a *(285)*
3. d *(285)*
4. a *(286)*
5. d *(284)*
6. b *(287)*
7. a *(286)*
8. b *(287)*
9. b *(287)*
10. c *(287)*
11. c *(288)*
12. b *(288)*
13. c *(288)*
14. d *(288, 289)*
15. d *(289)*
16. a *(289)*
17. c *(289)*
18. b *(289, 290)*
19. c *(290)*
20. c *(291)*
21. d *(291)*
22. d *(291)*
23. a *(291)*
24. c *(291)*
25. b *(292)*
26. c *(292)*
27. a *(293)*
28. b *(294)*
29. a *(294)*
30. c *(294)*
31. d *(294)*
32. b *(294, 308, 309)*
33. a *(295)*
34. d *(295)*
35. d *(296)*
36. c *(296)*
37. b *(297)*
38. c *(298)*
39. a *(298)*
40. c *(299)*
41. d *(300)*
42. a *(300)*
43. c *(300)*
44. b *(301)*
45. b *(301)*
46. a *(302)*
47. d *(302, 335)*
48. c *(302)*
49. a *(303)*
50. b *(304)*
51. c *(304)*
52. d *(304)*
53. b *(305, 306)*
54. b *(306)*

Identify

E.

1. Roof ladder *(285)*
2. Extension ladder *(286)*
3. Combination ladder *(287)*
4. Single (wall) ladder *(285)*
5. Folding ladder *(286)*
6. Pole (Bangor) ladder *(286)*
7. Pompier ladder *(287)*

Label

F.

a. Fly section *(282)*
b. Tip (Top) *(284)*
c. Rung *(284)*
d. Beam (Rail) *(281)*
e. Halyard *(283)*
f. Bed section *(281)*
g. Staypoles *(286)*
h. Butt spurs (Heel spurs) or Butt *(282)*
i. Pawl (Lock or Dog) *(283)*
j. Pulley *(283)*
k. Stop *(284)*

Chapter 10 Answers

FIREFIGHTER I

Matching

A.

1. b *(345)*
2. g *(345)*
3. a *(346)*
4. f *(347)*
5. e *(348)*
6. d *(364)*
7. i *(367)*
8. h *(368)*

B.

1. c *(361, **O&T** glossary)*
2. a *(363, **O&T** glossary)*
3. e *(366, **O&T** glossary)*
4. d *(367, **O&T** glossary)*
5. g *(368, **O&T** glossary)*
6. b *(370, **O&T** glossary)*

True/False

C.

1. True *(345)*
2. False. Insulation installed over roof coverings of fire-rated roof construction effectively retains heat and *may reduce the fire rating drastically,* causing premature roof failure. *(345)*
3. True *(346)*
4. False. Using stairwells or elevator shafts for evacuation and ventilation simultaneously *is potentially life-threatening. (352)*
5. False. Work with the wind at your back *or side* while cutting the roof opening. *(356)*
6. False. *Firefighters should evacuate the roof promptly* when ventilation work is complete. *(356)*
7. False. Ensure that *all personnel* on the roof are wearing full personal protective equipment *including SCBA. (356)*
8. True *(357)*
9. False. Almost every roof opening *will be locked or secured in some manner. (357)*
10. False. *Typically it is quicker to open an existing roof opening* than it is to cut a hole in the roof. *(358)*
11. True *(359)*
12. False. *Some slate and tile roofs may require no cutting.* Slate and tile roofs can be opened by using a large sledgehammer. *(359)*
13. False. When a significant amount of fire exists in the truss area of a roof structure, *firefighters should not be on or under a truss roof. (360)*
14. True *(360)*
15. False. Lightweight concrete slabs can be penetrated with a *hammerhead pick, power saw with concrete blade, jackhammer, or any other penetrating tool. (361)*
16. False. Operating a fire stream through a ventilation hole during offensive operations *stops the ventilation process and places interior crews in serious danger. (363)*
17. True *(364)*
18. False. Because horizontal ventilation is not accomplished at the highest point of a building, *there is constant danger that when the rising heated gases are released, they will ignite higher portions of the fire building. (365)*
19. True *(367)*
20. False. Forced-air fans should always be equipped with explosion-proof motors and power cable connections *when used in a flammable atmosphere. (367)*
21. True *(368)*

22. False. When using positive pressure to remove smoke from multiple floors of a building, it is *generally best to apply positive pressure at the lowest point.* *(368)*
23. True *(371)*

Multiple Choice

D.

1. d *(345)*
2. b *(345)*
3. a *(347)*
4. c *(348)*
5. b *(348, 349)*
6. d *(349)*
7. a *(350)*
8. c *(351)*
9. b *(354)*
10. c *(355)*
11. d *(356)*
12. a *(356)*
13. b *(356)*
14. a *(357)*
15. b *(357)*
16. c *(357)*
17. c *(357)*
18. d *(357)*
19. a *(358)*
20. b *(359)*
21. a *(359, 360)*
22. c *(360)*
23. b *(360)*
24. d *(361)*
25. c *(361)*
26. a *(361)*
27. d *(362)*
28. b *(363)*
29. a *(365)*
30. c *(367)*
31. b *(367)*
32. d *(367)*
33. d *(368)*
34. a *(368)*
35. c *(369)*
36. a *(370)*
37. b *(371)*
38. b *(370)*

Identify

E.

1. Flat *(359)*
2. Pitched *(359)*
3. Arched *(360)*
4. Trussless arched *(360)*
5. Concrete *(360)*
6. Corrugated galvanized sheet-metal *(361)*

Chapter 11 Answers

FIREFIGHTER I

Matching

A.

1. c *(379, dictionary)*
2. b *(390)*
3. e *(382)*
4. g *(382)*
5. a *(382)*
6. d *(386)*
7. f *(387)*
8. h *(391)*

B.

1. b *(384)*
2. c *(384)*
3. e *(384)*
4. a *(384)*
5. f *(384, 385)*

True/False

C.

1. False. *Water department officials should realize that fire departments are vitally concerned with water supply and work with them* on water supply needs and the locations and types of fire hydrants. *(379)*
2. False. In small towns, the *requirements for fire protection exceed other requirements. (379, 380)*
3. True *(380)*
4. False. Most communities use a combination of the direct pumping *and gravity systems. (381)*
5. True *(382)*
6. True *(383)*
7. False. Secondary feeders *should be arranged in loops as far as possible* to give two directions of supply to any point. *(383)*
8. False. The function of a valve in a water distribution system *is to provide a means for controlling the flow of water through the distribution piping. (383)*
9. False. A well-run water utility *has records of the locations of all valves. (384)*
10. True *(384)*
11. False. Nonindicating valves in water distribution systems are normally *buried or installed in utility chambers. (384)*
12. True *(385)*
13. True *(385)*
14. False. *All hard suction lines should have strainers* whenever drafting from a natural source. *(390)*
15. False. Plain siphons or *commercial tank-connecting devices are not generally as efficient as jet siphons. (391)*
16. True *(392)*
17. False. *Gravity dumps may be activated by a firefighter,* which relieves the driver/operator from exiting the cab and saves time in the overall process. *(393)*
18. False. The apparatus with the greatest pumping capacity *should be located at the water source. (393)*

Multiple Choice

D.

1. b *(379)*
2. a *(379)*
3. d *(379)*
4. b *(380, 381)*
5. c *(381)*
6. a *(382)*
7. b *(382, 383)*
8. d *(382, 383)*
9. c *(382, 383)*
10. b *(383)*
11. c *(383)*
12. a *(383)*
13. c *(383)*
14. b *(383)*
15. a *(384)*
16. d *(385)*
17. a *(385)*
18. d *(386, 387)*

19. b *(387)*
20. b *(388)*
21. c *(388)*
22. a *(388)*
23. c *(390)*
24. a *(390)*
25. d *(390)*
26. d *(391)*
27. c *(391)*
28. d *(391)*
29. a *(391)*
30. c *(391)*
31. d *(391)*
32. b *(393)*
33. c *(393)*

Identify

E.

1. Post indicator valve *(384)*
2. Outside screw and yoke *(384)*

F.

1. Post indicator valve *(384)*
2. Outside screw and yoke (OS&Y) valve *(384)*
3. Gate valve *(385)*
4. Butterfly valve *(385)*

G. *(383)*

1. Feeder
2. Secondary feeder
3. Distributors
4. Valves
5. Hydrants

FIREFIGHTER II

Matching

A.

1. a *(385)*
2. e *(385, 386)*
3. c *(386)*
4. d *(386)*
5. f *(386)*

True/False

B.

1. True *(382)*
2. False. Pressure in the fire service sense is measured in *pounds per square inch (psi) or kilopascal (kPa).* *(385)*
3. False. *Consideration should be given* to the effect that weather has on the amount of water available and the accesses to water sources. *(390)*

Multiple Choice

C.

1. d *(382)*
2. b *(382)*
3. a *(387)*
4. b *(388)*
5. b *(388)*
6. c *(388)*
7. d *(389)*
8. a *(390)*

Identify

D.

1. Pounds per square inch *(385)*
2. Kilopascal *(385)*

Chapter 12 Answers

FIREFIGHTER I

Matching

A.

1. g *(397)*
2. c *(397)*
3. a *(397)*
4. f *(397)*
5. e *(397)*
6. b *(397, 399)*

B.

1. c *(399)*
2. f *(403)*
3. a *(403)*
4. g *(404, 405)*
5. h *(405)*
6. d *(406)*
7. e *(406)*
8. k *(422)*
9. j *(415)*
10. i *(416)*

True/False

C.

1. False. References made to the diameter of fire hose refer to the dimensions of the *inside diameter* of the fire hose. *(397)*
2. True *(399)*
3. True *(399)*
4. False. Hose, *wet or dry,* should not be subjected to freezing conditions for prolonged periods of time. Hose can be damaged by freezing temperatures. *(400)*
5. False. The most commonly used couplings are the *threaded and Storz types. (402)*
6. False. Drop-forged couplings *stand up well to normal use. (402, 403)*
7. True *(403)*
8. False. Pin-lug couplings are not commonly ordered with new fire hose because of their *tendency to snag when hose is dragged over objects. (404)*
9. False. These couplings are designed to be connected and disconnected with only *one-third of a turn. (405)*
10. True *(405)*
11. False. *The swivel part should be submerged in a container of warm, soapy water and worked forward and backward* to thoroughly clean the swivel. *(405)*
12. False. *These two gaskets are not interchangeable.* The difference lies between their thickness and width. *(406)*
13. True *(414)*
14. True *(415)*
15. False. *Most hose beds have open slats in the bottom* that enable air to circulate throughout the hose load. *(416)*
16. True *(416)*
17. False. The horseshoe load *does not work well for large diameter hose* because the hose remaining in the bed tends to fall over as the hose pays off, which causes the hose to become entangled. *(417, 418)*
18. True *(418)*
19. False. Large diameter hose *can be loaded directly from the street or ground* by straddling the hose with the pumper and driving slowly forward as the hose is progressively loaded into the bed. *(418, 419)*
20. False. *Finishes for forward lays are usually designed to speed the pulling of a hose when making a hydrant connection* and are not as elaborate as finishes for reverse lays. *(419, 420)*
21. True *(422)*
22. True *(422)*
23. False. Threaded-coupling supply hose is usually arranged in the hose bed so that when hose is laid, *the female coupling is toward the water source and the male coupling is toward the fire. (422)*

24. False. Hose beds set up for forward lays should be loaded so that *the first coupling to come off the hose bed is female. (423)*
25. True *(424)*
26. False. With the reverse lay, *hose is laid from the fire to the water source. (425)*
27. False. The reverse lay has become a standard method for setting up a relay pumping operation *when using medium diameter hose as a supply line. (426)*
28. True *(425, 426)*
29. False. When reverse-laying a supply hose, *it is not necessary to use a four-way hydrant valve. (426)*
30. False. A hard suction hose *must be used when drafting from a static water source. (427)*
31. True *(427)*
32. True *(427)*
33. False. The working line drag is *one of the quickest and easiest ways to move fire hose at ground level, but it is limited by available personnel. (429)*
34. True *(430)*
35. False. Because advancing a hoseline down a stairway often subjects firefighters to intense heat, *the hoseline should be charged in most cases. (431)*
36. False. While hoselines may be pulled from the apparatus and extended to the fire area, *it is not considered good practice. It is more practical to have some hose rolled or folded on the apparatus ready for standpipe use. (431)*
37. False. If hose is already charged, *it is safer, quicker, and easier to drain the hose and relieve the pressure before advancement is made. (432)*
38. True *(433)*
39. False. Putting a kink in the hose *does not apply to LDH because of its size and weight when charged. (434)*
40. True *(435)*

Multiple Choice

D.

1. d *(397)*
2. c *(397)*
3. b *(397)*
4. b *(397)*
5. a *(397)*
6. a *(397)*
7. c *(397)*
8. d *(397)*
9. d *(399)*
10. c *(399, 400)*
11. a *(400)*
12. b *(400)*
13. d *(400)*
14. b *(400, 401)*
15. b *(400, 401)*
16. c *(401)*
17. a *(401)*
18. b *(401)*
19. a *(401)*
20. d *(402)*
21. c *(402)*
22. a *(403)*
23. b *(403)*
24. d *(403)*
25. c *(404)*
26. c *(405)*
27. a *(405)*
28. b *(406)*
29. d *(414)*
30. d *(414)*
31. a *(416)*
32. d *(417)*
33. c *(418)*
34. c *(419)*
35. b *(420)*
36. a *(420)*
37. d *(422)*
38. a *(423)*
39. a *(423)*
40. b *(424)*
41. c *(424)*
42. d *(424)*
43. c *(424)*
44. b *(426)*
45. d *(426)*
46. a *(428)*
47. c *(428)*
48. b *(429)*
49. c *(429)*
50. a *(429)*
51. a *(429)*
52. b *(430)*
53. a *(431)*
54. d *(432)*
55. c *(432, 433)*
56. b *(433)*
57. a *(434)*
58. d *(434)*
59. d *(434)*
60. a *(435)*
61. b *(434, 435)*
62. d *(436)*

E.

1. Straight roll *(414)*
2. Donut roll *(414)*
3. Twin donut roll *(415)*
4. Self-locking twin donut roll *(415)*

F.

1. Accordion load *(417)*
2. Horseshoe load *(418)*
3. Flat load *(418)*
4. Straight finish *(420)*
5. Reverse horseshoe finish *(420)*
6. Preconnected flat load *(421)*
7. Triple layer load *(421)*
8. Minuteman load *(422)*

G.

1. Threaded coupling *(402)*
2. Storz coupling *(402)*
3. Snap coupling *(403)*
4. Three-piece coupling *(403)*
5. Five-piece coupling *(403)*

FIREFIGHTER II

Matching

A.

1. b *(406)*
2. a *(409)*
3. g *(410)*
4. c *(410)*
5. i *(411)*
6. e *(412)*
7. h *(412)*
8. j *(412)*
9. d *(412)*
10. f *(413)*

B.

1. b *(406)*
2. d *(406)*
3. e *(406)*
4. a *(406)*

C.

1. c *(409)*
2. a *(409)*
3. f *(410)*
4. e *(410)*
5. d *(410)*

True/False

D.

1. False. Hose *appliances have water running through them* and tools do not. *(406)*
2. True *(406)*
3. False. *Wye appliances divide* a line of hose into two or more lines. *Siamese* fire hose layouts consist of two or more *hoselines that are brought into one hoseline or device. (406, 407)*
4. False. With the increased popularity of large diameter hose, siamese appliances are being used to *feed a large diameter hoseline when multiple smaller hoselines have to be used in the same relay as larger diameter hose. (408)*
5. True *(408)*
6. False. *By using the hydrant valve,* additional hoselines may be laid to the hydrant, *the supply pumper may connect to the hydrant,* and the pressure may be boosted in the original supply line *without having to interrupt the flow in the original supply line. (409)*
7. True *(409)*
8. False. Extending a line with a reducer *limits options to just that hoseline* whereas *using a gated wye at that point allows the option of adding another line if needed. (409)*
9. False. The hose jacket encloses the hose so effectively that *it can continue to operate at full pressure. (411)*
10. False. Firefighters should *never stand over the handle of a hose clamp when applying or releasing it. (411)*
11. True *(413)*
12. False. Hose should be tested in a place that has adequate room to *lay out the hose in straight runs, free of kinks and twists. (437)*
13. True *(437)*

Multiple Choice

E.

1. b *(406)*
2. a *(407)*
3. a *(407, 408)*
4. d *(408)*
5. b *(408)*
6. a *(408)*
7. d *(410)*
8. c *(411)*
9. d *(411)*
10. c *(411)*
11. a *(412)*
12. b *(437)*
13. d *(437)*
14. b *(437)*
15. a *(437)*
16. d *(437)*

F.

1. Ball valve *(407)*
2. Gate valve *(407)*
3. Butterfly valve *(407)*
4. Clapper valve *(407)*

G.

1. Wye appliance *(407)*
2. Siamese appliance *(407)*
3. Water thief appliance *(408)*
4. Forestry water thief appliance *(408)*
5. Manifold or LDH appliance *(408)*

Chapter 13 Answers

FIREFIGHTER I

Matching

A.

1. g *(487)*
2. b *(490)*
3. j *(491)*
4. d *(491)*
5. e *(491)*
6. a *(492)*
7. c *(493)*
8. h *(493)*
9. f *(494)*

True/False

B.

1. False. Complete vaporization does not happen the instant water reaches its boiling point *because additional heat is required to completely turn the water into steam. (487)*
2. False. Steam expansion is not gradual, *but rapid. (488)*
3. True *(489)*
4. False. When a nozzle is above the fire pump, *there is a pressure loss.* When the nozzle is below the pump, there is a pressure gain. *(490)*
5. False. *Operate* nozzle controls, hydrants, valves, and hose clamps *slowly to prevent water hammer. (491)*
6. True *(491, 492)*
7. True *(492)*
8. False. *Do not use solid streams on energized electrical equipment.* They can be conductors. *(494)*
9. False. *A straight stream is a pattern of the adjustable fog nozzle,* whereas a solid stream is discharged from a smoothbore nozzle. *(494)*
10. True *(495)*
11. False. Shorter reach is why *fog streams are seldom useful for outside,* defensive fire fighting operations. *(495)*
12. False. The firefighter has the choice of making flow-rate adjustments *either before opening the nozzle or while water is flowing. (495, 496)*
13. True *(496)*
14. True *(496)*
15. False. Because a broken stream may have sufficient continuity to conduct electricity, *it is not recommended for use on Class C fires. (497)*
16. False. The ball can be rotated up to *90 degrees. (497)*
17. False. While it *will operate in any position between fully closed and fully open,* operating the nozzle with the valve in the fully open position gives maximum flow and performance. *(497)*

Multiple Choice

C.

1. d *(488)*
2. a *(489)*
3. c *(490)*
4. b *(490)*
5. c *(491)*
6. a *(491, 492)*
7. d *(492)*
8. b *(491)*
9. b *(493)*
10. d *(493)*
11. a *(494)*
12. d *(494)*
13. c *(494)*
14. a *(496)*
15. d *(492, 495)*
16. b *(495)*
17. d *(496)*
18. c *(496)*
19. b *(496)*
20. a *(496)*
21. c *(496, 497)*
22. d *(498)*
23. a *(497)*
24. c *(497)*
25. b *(498)*

Identify

D.

1. Gallons per minute *(491, **O&T** glossary)*
2. Liters per minute *(491, **O&T** glossary)*

E.

1. Ball valve *(497)*
2. Slide valve *(497)*
3. Rotary control valve *(498)*

FIREFIGHTER II

Matching

A.

1. a *(499)*
2. e *(499)*
3. d *(499)*
4. f *(499)*
5. c *(499)*

True/False

B.

1. True *(499)*
2. False. Proper aeration produces *uniform-sized foam bubbles* to provide a longer lasting blanket. *(499)*
3. False. *Failure to match the proper foam concentrate with the burning fuel* will result in an unsuccessful extinguishing attempt and could endanger firefighters. *(500)*
4. False. *Reducing surface tension* provides for better water penetration, thereby increasing its effectiveness. *(500)*
5. True *(505)*
6. False. Stopping and restarting *may allow the fire and fuel to consume whatever foam blanket has been established. (505)*
7. True *(505)*
8. False. *The proportioning percentage for Class A foams can be adjusted* (within limits recommended by the manufacturcr) to achieve specific objectives. *(505)*
9. False. Using a foam proportioner that is not compatible with the delivery device *(even if the two are made by the same manufacturer)* can result in unsatisfactory foam or no foam at all. *(507)*
10. True *(508)*
11. True *(508)*
12. False. Portable foam proportioners are the *simplest and most common foam proportioning devices in use today. (510)*
13. True *(510)*
14. False. These nozzles may be used with alcohol-resistant AFFF foams on hydrocarbon fires *but should not be used on polar solvent fires. (511, 512)*
15. False. Air-aspirating foam nozzles provide maximum expansion of the agent, but *the reach of the stream is less than that of a standard fog nozzle. (512)*
16. False. Foam concentrates, either *at full strengths* or in diluted forms, *pose minimal health risks to firefighters. (515)*
17. True *(515)*
18. False. The *less oxygen required* to degrade a particular foam, the better or more environmentally friendly the foam is when it enters a body of water. *(515)*

Multiple Choice

C.

1. c *(498)*
2. b *(498)*
3. a *(498)*
4. d *(498, 499)*
5. b *(499)*
6. c *(500)*
7. d *(500)*
8. b *(500)*
9. a *(500)*
10. c *(500)*
11. d *(505)*
12. d *(505)*
13. c *(508)*
14. b *(508)*
15. d *(508)*
16. d *(508)*
17. a *(508, 509)*
18. a *(509)*
19. b *(510)*
20. b *(510)*
21. a *(510)*
22. c *(511)*
23. b *(511)*
24. d *(511, 512)*
25. a *(513)*
26. d *(513)*
27. c *(513)*
28. b *(514)*
29. d *(514)*
30. a *(514)*
31. c *(515)*

Identify

D.

1. Compressed air foam systems *(500)*
2. Aqueous film forming foam *(504)*
3. Film forming fluoroprotein foam *(504)*

4. Aircraft rescue and fire fighting *(511)*
5. Material safety data sheet *(515)*

E.

1. In-line eductor *(510)*
2. Foam nozzle eductor *(511)*

F.

1. Solid bore nozzle *(512)*
2. Fog nozzle *(512)*

Chapter 14 Answers

FIREFIGHTER I

Matching

A.

1. d *(529)*
2. b *(536, dictionary)*
3. e *(537)*
4. c *(540)*

B.

1. f *(554)*
2. c *(554)*
3. i *(554)*
4. h *(554)*
5. a *(554)*
6. e *(555)*
7. g *(555)*
8. j *(555)*
9. d *(555)*

C.

1. a *(554)*
2. f *(554)*
3. g *(554)*
4. e *(554)*
5. d *(556)*
6. c *(556)*

True/False

D.

1. False. *Depending on the conditions at the fire scene,* the fire officer may choose to perform immediate rescue or to protect exposures rather than attack the fire. *(522)*
2. False. Any burning fascia and soffit, boxed cornices, or other doorway overhangs *should be extinguished before entry. (523)*
3. True *(525)*
4. False. *Water should not be applied for too long a time,* otherwise thermal layering will be upset. *(525)*
5. False. *An indirect attack is not desirable where victims may yet be trapped* or where the spread of fire to uninvolved areas cannot be contained. *(527)*
6. True *(528)*
7. False. The primary danger of electrical fires is *the failure of emergency personnel to recognize the safety hazard. (535)*
8. True *(535)*
9. True *(535)*
10. False. *Fires in transformers can present a serious health and environmental risk* because of coolant liquids that contain PCBs. *(536)*
11. False. *It is no longer recommended to pull the electrical meter* to turn off the electricity in residential fires. *(538)*
12. True *(539)*
13. False. The control box should be *tagged to indicate that it is out of service. (539)*
14. False. *Consider all downed electrical wires equally dangerous* even when one is arcing and others are not. *(540)*
15. True *(540)*
16. False. Firefighters should not assume that these fires are extinguished *just because flames are not visible.* Combustible metal fires are very hot — greater than 2,000°F *(1 093°C)* — even if they appear suppressed. *(541)*
17. True *(541)*
18. False. Initially, the truck company *observes the outside of the building for signs of victims needing rescue. (543)*
19. False. In most engine compartment fires, *the fire must be knocked down before the hood can be opened. (548)*
20. False. Firefighters should *not assume that any vehicle is without extraordinary hazards. (549)*

21 True *(553)*

22. True *(554)*
23. False. Wildland fires will *usually move faster uphill* than downhill. *(554)*

FIREFIGHTER I

24. False. The green is the area of unburned fuels next to the involved area. The green *does not indicate a safe area. (555)*
25. True *(556)*
26. True *(556)*

Multiple Choice

E.

1. b *(522)*
2. c *(523)*
3. a *(523)*
4. d *(524)*
5. c *(524)*
6. a *(525)*
7. b *(526)*
8. d *(527)*
9. c *(527)*
10. a *(528)*
11. b *(528)*
12. c *(528)*
13. d *(528, 529)*
14. b *(529)*
15. d *(529)*
16. c *(535)*
17. a *(535)*
18. b *(536)*
19. c *(536)*
20. d *(536)*
21. b *(536)*
22. d *(537)*
23. a *(537)*
24. c *(538)*
25. a *(539)*
26. b *(540)*
27. b *(541)*
28. a *(541)*
29. d *(543)*
30. c *(548)*
31. c *(549)*
32. b *(552)*
33. b *(553)*
34. a *(553, 554)*

Identify

F.

1. Liquefied petroleum gas *(529)*
2. Polychlorinated biphenyl *(536)*
3. Supplemental Restraint System *(548)*
4. Side-Impact Protection System *(548)*

G. *(556)*

1. **F**ight fire aggressively but provide for safety first.
2. **I**nitiate all action based on current and expected fire behavior.
3. **R**ecognize current weather conditions and obtain forecasts.
4. **E**nsure instructions are given and understood.
5. **O**btain current information on fire status.
6. **R**emain in communication with crew members, your supervisor, and adjoining forces.
7. **D**etermine safety zones and escape routes.
8. **E**stablish lookouts in potentially hazardous situations.
9. **R**etain control at all times.
10. **S**tay alert, keep calm, think clearly, act decisively.

FIREFIGHTER II

Matching

A.

1. e *(529)*
2. b *(529)*
3. a *(529)*
4. d *(530)*
5. c *(534)*

True/False

B.

1. True *(530)*
2. False. *Unless the leaking product can be turned off,* fires around relief valves or piping *should not be extinguished.* Contain the pooling liquid until the flow is stopped. *(530)*
3. False. Firefighters *should not assume that relief valves are sufficient to safely relieve excess pressures* under severe fire conditions. Firefighters have been killed by the rupture of both large and small flammable liquid vessels that have been subjected to flame impingement. *(530)*
4. False. Experience has shown that water in various ways is *effective in extinguishing or controlling many Class B fires.* Water will be most useful as a cooling agent for protecting structures. *(531)*
5. True *(531)*
6. True *(531)*
7. False. Using hoselines as a protective cover *must be practiced through training before being attempted during an emergency. (532)*
8. False. *Never approach horizontal vessels exposed to fire from the ends* because of the danger of ruptures; vessels frequently split and become projectiles. *(532)*
9. False. Natural gas in a compressed state *is subject to BLEVE. (533)*
10. True *(534)*
11. False. *Firefighters should not attempt to operate main valves* because incorrect action may worsen the situation or cause unnecessary loss of service to areas unaffected by the break. *(534)*
12. True *(544)*
13. True *(544)*
14. False. Once firefighters reach the basement, *the heat conditions are similar to those of a structural fire attack. (545)*
15. True *(547)*
16. False. Firefighters should *not enter these enclosures until the incident commander has decided on a course of action and issued specific orders. (550)*

Multiple Choice

C.

1. d *(530)*
2. b *(530)*
3. a *(531)*
4. a *(531)*
5. c *(531)*
6. d *(532, 533)*
7. d *(533)*
8. b *(533)*
9. d *(533)*
10. a *(534)*
11. d *(534)*
12. d *(534)*
13. c *(534)*
14. c *(544)*
15. d *(545)*
16. a *(545)*
17. c *(547)*
18. d *(550)*
19. b *(550)*
20. a *(551)*
21. c *(551)*

Identify

D.

1. Boiling liquid expanding vapor explosion *(530)*
2. Liquefied petroleum gas *(533)*
3. Compressed natural gas *(533)*
4. Rapid intervention crew *(544)*
5. Parts per million *(551)*

Chapter 15 Answers

FIREFIGHTER I

True/False

A.

1. False. *An early arriving pumper should connect to the FDC* in accordance with the pre-incident plan. *(583)*
2. True *(584)*
3. False. For liability purposes, it is *not recommended that fire department personnel service system components. (584)*
4. True *(584)*
5. False. Sprinkler system control valves *must be open for proper operation;* firefighters should check to see that they are. *(583)*

Identify

B. *(573)*
C

FIREFIGHTER I

FIREFIGHTER II

Matching

A.

1. e *(559)*
2. b *(560)*
3. j *(560)*
4. a *(561)*
5. h *(561)*
6. d *(562)*
7. g *(563)*
8. c *(567)*
9. i *(568)*

B.

1. f *(572)*
2. d *(572)*
3. a *(572)*
4. b *(573)*
5. c *(577)*
6. g *(580)*

C.

1. d *(580)*
2. b *(580)*
3. a *(580)*
4. e *(581)*
5. c *(581, 582)*

True/False

D.

1. True *(559)*
2. False. While detectors of this type are still in service, *their manufacture has been discontinued. (560, 561)*
3. False. Most bimetallic detectors *will reset automatically* when cooled. *(562)*
4. False. A smoke detector does not have to wait for heat to be generated and *can initiate an alarm of fire much more quickly than a heat detector.* For this reason, *smoke detectors are the preferred detectors in many types of occupancies. (563)*
5. True *(564)*
6. False. The *decrease in the current* flowing between the plates initiates an alarm signal. *(564)*
7. False. In some rural areas and areas with high thunderstorm occurrence, power failures may be frequent, and *battery-operated units may be more appropriate. (565)*
8. True *(568)*
9. True *(570)*
10. False. These systems represent another generation of technology that *continues to provide reliable protection as long as the systems are tested regularly and maintained in accordance with the manufacturer's instructions. (570)*
11. True *(572)*
12. False. Sprinklers *cannot be interchanged* because they are not designed to provide the proper spray pattern and coverage in any other position. *(575)*
13. False. Control valves are located *between the source of water supply and the sprinkler system. (576)*
14. False. The main control valve should always be returned to the *open position* after maintenance is complete. *(576, 577)*
15. True *(579)*
16. False. The deluge system is designed to operate automatically, and the sprinklers *do not have heat-responsive elements; therefore, it is necessary to provide a separate detection system. (582)*

Multiple Choice

E.

1. d *(560)*
2. b *(561)*
3. d *(562)*
4. a *(562)*
5. d *(562)*
6. c *(562)*
7. a *(563)*
8. b *(565)*
9. c *(565)*
10. a *(566)*
11. d *(567)*
12. c *(570)*
13. b *(579)*
14. a *(571)*
15. b *(572)*
16. b *(574)*
17. c *(574)*
18. c *(575)*
19. b *(575)*
20. a *(575)*
21. d *(576)*
22. b *(578)*
23. b *(578)*
24. d *(579)*
25. a *(579)*
26. d *(581)*
27. d *(581)*
28. c *(582)*
29. a *(582)*
30. c *(582, 583)*

Identify

F.

1. Ultraviolet *(565)*
2. Infrared *(565)*
3. Americans with Disabilities Act *(566)*
4. Outside screw and yoke *(577)*
5. Post indicator valve assembly *(577)*
6. Post indicator valve *(577)*
7. Wall post indicator valve *(577)*
8. Fire department connection *(572)*

G.

1. Bimetallic *(562)*
2. Rate-of-rise spot heat *(562)*
3. Pneumatic rate-of-rise heat *(563)*
4. Rate-compensated heat *(563)*
5. Thermoelectric heat *(563)*
6. Fire-gas *(566)*
7. Combination heat/smoke *(566)*

H.

1. Fusible-link *(574)*
2. Frangible-bulb *(574)*
3. Chemical-pellet *(574)*
4. Pendant *(575)*
5. Upright *(575)*
6. Sidewall *(575)*

I.

1. Outside screw and yoke valve *(577)*
2. Post indicator valve *(577)*
3. Wall post indicator valve *(577)*
4. Post indicator valve assembly *(577)*
5. Quarter-turn alarm test valve *(577)*
6. Inspector's test valve *(578)*

Label

J. *(573)*

1. Cross main
2. Branch lines
3. Pipe hangers
4. Automatic sprinklers
5. Inspector's test valve
6. Water supply
7. Main drain valve
8. Fire department connection
9. Water flow alarm
10. Riser
11. Alarm check valve
12. Main control valve (OS&Y)

K. *(573)*

1. Deflector
2. Fusible link
3. Lever arms
4. Valve cap
5. Toggle joint
6. Frame arms

Chapter 16 Answers

FIREFIGHTER I

Matching

A.

1. e *(587)*
2. a *(587)*
3. c *(587)*
4. h *(592)*
5. d *(592)*
6. g *(593)*
7. f *(593)*

True/False

B.

1. False. Salvage starts *as soon as adequate personnel are available* and may be done simultaneously with fire attack. *(587)*
2. False. Overhaul operations are *not normally started until the fire is under control. (587)*
3. True *(588)*
4. False. It is *very difficult to adapt plastic covers to traditional salvage cover folds. (590)*
5. False. Canvas salvage covers should be *completely dry* before they are folded and placed in service. *(590, 591)*
6. True *(591)*
7. False. Fire department pumpers *should not be used for removing water from basements, elevator shafts, or sumps* because pumpers are intricate and expensive machines that are not intended to pump the dirty, gritty water found in such places. *(593)*
8. False. A salvage cover rolled for a one-firefighter spread *may be carried on the shoulder* or under the arm. *(594)*
9. True *(594)*
10. False. *Plastic sheeting,* a hammer-type stapler, and duct tape *allow quick and easy construction of water diversion chutes. (595)*
11. True *(596)*
12. False. Many of the tools and equipment used for overhaul *are the same as those used for other fire fighting operations. (596)*
13. False. Typically, overhaul *begins in the area of actual fire involvement. (597)*
14. False. *Usually, it is unnecessary to remove insulation material* in order to properly check it or extinguish fire in it. *(598)*
15. False. *Only enough wall, ceiling, or floor covering should be removed* to ensure complete extinguishment. *(598)*
16. True *(598)*

Multiple Choice

C.

1. d *(587)*
2. a *(588)*
3. b *(588, 589)*
4. c *(590)*
5. a *(590)*
6. d *(590)*
7. d *(591)*
8. c *(591)*
9. b *(592)*
10. a *(594)*
11. d *(595)*
12. c *(596)*
13. b *(596)*
14. d *(596)*
15. c *(596)*
16. a *(597)*
17. b *(599)*
18. c *(599)*

Chapter 17 Answers

FIREFIGHTER I

True/False

A.

1. False. The important point is *not when* the firefighter notices something that can lead to the fire cause but that the firefighter *takes the proper steps afterwards. (623)*
2. False. *Each firefighter* should write a chronological account of important circumstances *personally* observed. *(627)*
3. False. Report hearsay to the investigator for validation. *It may be very helpful to the investigator. (627)*
4. True *(627)*
5. False. Firefighters *should not gather or handle evidence* unless it is absolutely necessary in order to preserve it. *(629)*
6. True *(629)*
7. False. *After evidence has been properly collected by an investigator,* debris may be removed. *(629)*

Multiple Choice

B.

1. a *(623)*
2. c *(627)*
3. d *(629)*
4. c *(629)*
5. a *(629)*

Case Study

C. *(623–627)*

Answers can be in question or statement format.

1. Why was a man at the construction site at 2 a.m.?

   Why was the man dressed in work clothes in the middle of the night?

   Why was the building burning in two separate areas?

2. Why was a vehicle blocking the fire hydrant?

   Why was the garage door open? Is there evidence that entry was forced through the garage door?

   Why was the space heater set up near the workbench during late summer?

3. What were the makes, models, colors, and license tag numbers of the vehicles leaving the parking lot?

   Why was only one of the loading dock doors opened?

   Why were objects blocking entry into the office area?

   Why was the owner so eager to find the insurance policy?

4. Was the old furniture placed in the hallway to hinder the fire attack?

   Were the rekindles the result of an accelerant used to start the fire?

   Was the broken glass a remnant of an incendiary device?

   Why did the fire alarm fail to work? Could the cause be human intervention?

5. Was the path of charred carpet the result of a trailer?

   Could the stack of papers on the bed have been important or incriminating documents?

   Could someone have tampered with the hydrant?

FIREFIGHTER II

True/False

A.

1. True *(621)*
2. False. Investigators *are seldom present* while firefighters fight a fire, perform overhaul, and interview occupants and witnesses to obtain information. *(621)*
3. False. The moment one suspects a particular person of arson, *the firefighter should call a trained investigator to conduct an interview. (627)*
4. True *(628)*
5. False. There are *no specific boundaries* for the cordon. *(628)*
6. True *(628)*
7. False. *Firefighters may seize [without a warrant] evidence of arson that is in plain view.* In many instances a search warrant or written consent to search will be needed for further visits to the premises. *(628)*
8. True *(629)*

Multiple Choice

B.

1. a *(621)*
2. d *(622)*
3. b *(622)*
4. c *(627)*
5. d *(627)*
6. a *(628)*
7. b *(628)*
8. d *(629)*

Chapter 18 Answers

FIREFIGHTER I

True/False

A.

1. False. A telecommunicator *must know* where emergency resources are in relation to the reported incident as well as their availability status. *(634)*
2. False. The worthiness of individuals for assistance *should not be evaluated by the telecommunicator. (634, 635)*
3. False. In an age of computers and electronic displays, *it is still vital for a telecommunicator to be able to look at a map and locate specific points. (635)*
4. False. Instead of reducing the need for using maps, *the opposite is the case with this technology. (636)*
5. True *(637, 638)*
6. False. A computer-generated document does *not need to be converted to hard copy*—the file is digitized and sent over the transmission medium. *(638)*
7. True *(642)*
8. False. Extremely loud audible devices or bright lights that come on in the middle of the night *can be somewhat shocking to a firefighter and can raise anxiety levels. (644, 645)*
9. True *(645)*
10. True *(645)*
11. True *(645)*
12. False. *It is a federal offense to send personal or nonemergency service messages* over a designated fire department radio channel. *(645)*
13. False. The company officer *does not provide the reason or justification* for requests made over the air. *(646)*
14. True *(648)*

Multiple Choice

B.

1. a *(633)*
2. d *(634)*
3. c *(634)*
4. a *(635)*
5. b *(636)*
6. d *(637)*
7. c *(638)*
8. b *(638)*
9. a *(639)*
10. c *(639)*
11. c *(641)*
12. d *(641)*
13. b *(641)*
14. a *(641)*
15. c *(641, 642)*
16. d *(642)*
17. b *(642)*
18. c *(642)*
19. d *(644)*
20. a *(646)*
21. d *(646)*
22. b *(647)*
23. c *(647)*
24. a *(648)*
25. b *(648)*
26. c *(648)*
27. d *(649)*
28. b *(649)*

Identify

C.

1. Computer-Aided Dispatch *(635)*
2. Automatic Vehicle Locating system *(636)*
3. Telecommunications Device for the Deaf *(637)*
4. Teletype *(637)*
5. Automatic Number Identification *(641)*
6. Automatic Location Identification *(641)*
7. Citizens band *(642)*
8. Federal Communications Commission *(645)*

FIREFIGHTER II

True/False

A.

1. True *(649)*
2. False. Incident reports are *available to the public. (651)*
3. False. Incident reports can also be *directly entered into a computer* by the officer in charge. *(651)*

Multiple Choice

B.

1. a *(649)*
2. c *(649)*
3. c *(651)*
4. a *(651)*

Identify

C.

1. United States Fire Administration *(651)*
2. National Fire Incident Reporting System *(651)*
3. Personal computer *(651)*

D. *(651)*

1, 3, 5, 6, 7, 8

Chapter 19 Answers

FIREFIGHTER I

Matching

A.

1. d *(656)*
2. e *(656)*
3. c *(656)*
4. b *(656)*

B.

1. a *(658)*
2. d *(658)*
3. c *(658)*
4. e *(659)*

True/False

C.

1. True *(660)*
2. False. Fire safety surveys in existing residential occupancies can only be accomplished *on a voluntary basis.* *(666)*
3. False. *Educating citizens, at all ages,* to recognize potential hazards and take appropriate action is a fire department function. *(671)*
4. True *(671)*
5. False. Firefighters must *skillfully sell the audience on the value* of making safe practices a way of life. *(672)*
6. True *(672)*
7. False. The test buttons on some detectors *may only check the device's horn circuit* so being cognizant of the manufacturer's smoke-test procedure is vital to maintaining a functional detector. *(675)*
8. True *(675)*
9. False. *Never* allow visitors, *especially young children,* to roam around the fire station unescorted. *(676)*

Multiple Choice

D.

1. d *(656)*
2. b *(658)*
3. c *(658)*
4. a *(658)*
5. d *(659)*
6. a *(660, 661)*
7. c *(666)*
8. d *(666)*
9. a *(667)*
10. b *(667)*
11. d *(668)*
12. c *(668)*
13. a *(673)*
14. d *(674)*
15. a *(676)*

Identify

E.

1. Public service announcement *(655)*
2. Exit drills in the home *(667)*
3. Factory Mutual *(669)*

F. *(672, 673)*

2, 4, 6, 8, 10

G. *(669, 670)*

1. How often is the chimney cleaned?

   Where is the wood for the fireplace stored?

   Stacks of magazines and newspapers are fire hazards.

   Smoke detectors are installed correctly and in all of the recommended locations.

   How often are the smoke detectors tested and the batteries changed?

   Rags soaked with cleaning agents should not be left near a hot water heater.

   Are all the household appliances in working order?

   How high is the hot water heater temperature set?

Are the dryer and exhaust vents cleaned regularly of lint?

Is the portable heater listed with UL, FM, or another testing laboratory?

2. Note the composition and condition of the roof.

   Are the spark arrestors in the chimney regularly checked?

   Window bars should be equipped with devices for opening them from the inside.

   The vegetation is properly maintained.

   Paint and brushes should not be stored under the porch.

   Outside waste burners are not recommended.

   The harmful chemicals are appropriately locked in a shed out of children's reach.

   Are the pesticides, fertilizer, and extra gasoline properly stored?

   Flammable liquids should be stored in safety-type cans.

FIREFIGHTER II

True/False

A.

1. False. An earnest effort by firefighters to create a favorable impression upon the owner helps to establish *a courteous and cooperative relationship. (662)*
2. False. A representative of the occupant should accompany firefighters during the entire survey. *Such a guide will help obtain ready access to all areas of the building. (662)*
3. True *(664)*
4. False. *It should be recommended* that a marking system such as that outlined in NFPA 704 be affixed to the outside of such structures. *(664)*
5. True *(664)*

Multiple Choice

B.

1. d *(662)*
2. a *(662, 663)*
3. c *(663)*
4. d *(664)*
5. b *(664)*
6. c *(665)*
7. a *(666)*
8. d *(666)*

Identify

C.

1. Hazardous materials *(664)*
2. Graphic Information System *(664)*

Label

D. *(665)*

Fire Protection

1. Fire Department Connection
2. Automatic Sprinklers throughout contiguous sections of single risk
3. Automatic Sprinklers all floors of building
4. Automatic Sprinklers in part of building only (Note under symbol indicates protected portion of building)
5. Not Sprinklered
6. Automatic Chemical Sprinklers
7. Chemical Sprinklers in part of building only (Note under symbol indicates protected portion of building)
8. Vertical Pipe or Standpipe
9. Automatic Fire Alarm
10. Water Tank
11. Fire Escape
12. Fire Alarm Box
13. Single Hydrant
14. Double Hydrant
15. Triple Hydrant
16. Quadruple Hydrant of the High Pressure Fire Service
17. Water Pipes of the High Pressure Fire Service
18. Water Pipes of the High Pressure Fire Service as Shown on Key Map
19. Public Water Service
20. Private Water Service
21. Fire Detection System — Label Type
22. Alarm Gong, with Hood
23. Sprinkler Riser (size indicated)

Vertical Openings

1. Skylight lighting top story only
2. Skylight lighting 3 stories
3. Skylight with Wired Glass in Metal Sash
4. Open Elevator
5. Frame Enclosed Elevator
6. Frame Enclosed Elevator with Traps
7. Frame Enclosed Elevator with Self-Closing Traps
8. Concrete Block Enclosed Elevator with Traps
9. Tile Enclosed Elevator with Self-Closing Traps
10. Brick Enclosed Elevator with Wired Glass Door
11. Open Hoist
12. Hoist with Traps
13. Open Hoist Basement to 1st
14. Stairs

Miscellaneous

1. Number of Stories
   Height in Feet
   Composition Roof Covering
2. Parapet 6″ above Roof
   Frame Cornice
   Parapet 12″ above Roof
3. Parapet 24″ above Roof
   Occupied by Warehouse
   Metal, Slate, Tile, or Asbestos
   Shingle Roof Covering
   Parapet 48″ above Roof
4. 2 Stories and Basement
   1st Floor Occupied by Store
   2 Residential Units above 1st
   Auto in Basement
   Drive or Passageway
   Wood Shingle Roof
5. Iron Chimney
6. Iron Chimney (with Spark Arrestor)
7. Vertical Steam Boiler
8. Horizontal Steam Boiler
9. Width of Street between Block Lines, not Curb Lines
   Ground Elevation
   House Numbers nearest to buildings are official or actually up on buildings. Old House Numbers are farthest from buildings.
10. Brick Chimney
11. Gasoline Tank
12. Fire Pump

Color Code for Construction Materials for Walls

1. Fire-Restrictive protected steel
2. Brick, hollow tile
3. Frame—wood, stucco
4. Concrete, stone, or hollow concrete block
5. Noncombustible unprotected steel